Reuse of sewage effluent

Proceedings of the international symposium
organized by the Institution of Civil Engineers
and held in London on 30–31 October 1984

Thomas Telford, London

Symposium sponsored jointly by the Institution of Civil Engineers, the International Commission on Irrigation and Drainage, and the Ross Institute of Tropical Hygiene.

Organizing Committee: F. G. Poskitt (Chairman), J. P. Cowan, Professor P. C. G. Isaac, G. W. Morrey, L. H. Thompson

British Library Cataloguing in Publication data

Reuse of sewage effluent
    1. Sewage — purification
    2. Water reuse
    I. Institution of Civil Engineers
    628.3   TD429

ISBN: 0 7277 0230 0

First published 1985

© Institution of Civil Engineers, 1984, 1985, unless otherwise stated

Published for the Institution of Civil Engineers by Thomas Telford Ltd, P. O. Box 101, 26–34 Old Street, London EC1P 1JH

Printed in Great Britain by Billing & Sons Ltd, Worcester.

Contents

Opening address   1

POTENTIAL NEED FOR REUSE
1. Reuse via rivers for water supply. D. D. YOUNG   11

2. The urgent need for the reuse of sewage effluent in developing countries. B. Z. DIAMANT   23

3. Reuse of sewage effluent in Australia. A. G. STROM   35

Discussion on Papers 1 – 3   45

REUSE FOR INDUSTRY
4. Reuse of industrial effluent. R. C. SQUIRES   53

5. Reuse of sewage effluent in industry. M. R. G. TAYLOR and J. M. DENNER   77

Discussion on Papers 4 and 5   89

REUSE FOR AGRICULTURE
6. Urban effluent reuse for agriculture in arid and semi-arid zones. M. B. PESCOD and U. ALKA   93

7. Reuse of effluent for agriculture in the Middle East. J. P. COWAN and P. R. JOHNSON   107

Discussion on Papers 6 and 7   129

FISH FARMING
8. Constructive uses of sewage with regard to fisheries. R. J. HUGGINS   147

9. Use of sewage waste in warm water aquaculture. A. I. PAYNE   157

REUSE FOR AMENITY
10. A Dutch example and continental practice. H. M. J. SCHELTINGA   173

11. Guidelines for evaluating recreational water reuse. J. F. CARUSO and R. J. AVENDT   183

Discussion on Papers 10 and 11   193

REUSE FOR RECHARGE
12. Cedar Creek reclamation — groundwater recharge demonstration program. R. J. AVENDT   199

14. Groundwater recharge of sewage effluents in the UK.
H. A. MONTGOMERY, M. J. BEARD and K. M. BAXTER   219

Discussion on Papers 12 and 14   227

HEALTH ASPECTS
15. Health aspects of wastewater reuse. R. G. FEACHEM and
D. BLUM   237

16. Reuse of sewage effluent in the UK — an appraisal of
health related matters. J. W. RIDGWAY   249

Discussion on Papers 15 and 16   261

PROJECT APPRAISAL AND ECONOMICS
17. Sewage effluent reuse — economic aspects in project
appraisal. R. B. PORTER and F. FISHER   267

Discussion on Paper 17   281

ADVANCED TREATMENT AND RESEARCH
18. Advanced treatment processes for the reclamation of
water from sewage effluent. C. BOWLER and
S. H. GREENHALGH   285

19. Process selection and design aspects relating to advanced
treatment of sewage effluents. L. R. J. VAN VUUREN and
B. M. VAN VLIET   299

Discussion on Papers 18 and 19   315

General discussion   323

Closing address   327

Addendum   330

Paper 13 was withdrawn

Opening address

Professor D. A. OKUN, Department of Environmental Sciences and Engineering,
The University of North Carolina, USA

INTRODUCTION

Water reuse has been practised since water was introduced for
the removal of household wastes. An early and still common
example is the discharge of wastewaters into rivers, whence,
after more or less dilution, they are abstracted for water
supply downstream. The consequences of this reuse in London
in the mid-19th century, with the frequent outbreaks of
cholera, marked the beginning of 'The great sanitary
awakening'. The disposal of wastewaters onto land,
accompanied by the growing of useful crops, the venerable so-
called 'sewage farm', which is still today being proposed and
adopted as a method for wastewater disposal, is often cited as
an example of reuse. Then why are symposia such as this and
Water Reuse Symposium III, held in San Diego, California, in
August 1984, eliciting new interest?

The reason is simple. We are now finding that effluent is
no longer merely the cause of nuisance to be abated but rather
a resource with value. Effluent is being sold both wholesale
and retail and now being fought over (in the law courts) in
the American West. The value of wastewater as a resource is
naturally expected to be high in arid and semi-arid areas of
the world, but we are finding reclaimed wastewater to be a
valuable resource in humid areas as well.

Before we discuss reuse, we need to distinguish between the
various types of reuse, so that we are clear as to the type of
reuse being discussed. I will define each type and then
concentrate on that which I believe has the greatest promise
for purposeful implementation.

Indirect potable reuse

The abstraction of water for drinking and other purposes from
a surface or underground source into which treated or
untreated wastewaters have been discharged.

Direct potable reuse

The piping of treated wastewaters directly into a water supply
system that provides water for drinking.

Indirect non-potable reuse
The abstraction of water for one or more non-potable purposes from a surface or underground source into which treated or untreated wastewaters have been discharged.

Direct non-potable reuse
The piping of treated wastewaters directly into a water supply system that provides water for one or more non-potable purposes.

Recycling
The use of wastewater within an establishment for a useful purpose, such as cooling water for process water within an industry. Recycling, although important in providing savings to industry and the community, is not considered in this paper.

The disposal of wastewaters on land, where the criteria for design and operation are based on the need for pollution abatement, has a place in the armamentarium of pollution control but is not here considered reuse.

INDIRECT POTABLE REUSE
This type of reuse is of long standing and it is ubiquitous. Because engineers and scientists at the turn of the century learned how to render waters containing human wastes safe to drink, primarily through filtration and disinfection with chlorine, the transmission of water-borne infectious diseases has virtually been eliminated in the industrialized countries, although these diseases continue to plague the peoples of Asia, Africa and Latin America. Unfortunately, we can no longer be sanguine about the safety of such indirect potable reuse because of the synthetic organic chemicals that are inevitably present in water supplies drawn from sources that receive wastewaters and runoff from urban and industrial areas.

Where only a few years ago drinking water standards in the US and in Europe did not contain any limits for specific synthetic organic chemicals, now they do. Furthermore, as we learn more about their health significance, new chemicals are being added to the list and the permissible concentrations of those already present are being reduced. Of particular concern is that each chemical is addressed as if it were the only contaminant present when we can assume that the impact of these chemicals is at least additive and may, in fact, be synergistic. A hint of what is ahead may be discerned from the health effects studies reported in the National Research Council (US) publication 'Drinking water and health' - the first volume of which was issued in 1977, with the fifth volume being published in 1983, and more to come.

Knowledge about the health effects of these chemicals and the technology for their monitoring and removal will always lag behind the development of new chemicals. They are introduced into commerce and industry at the rate of some 1000 annually and ultimately find their way into water courses that drain urban and industrial areas. Accordingly, while we may need to make the best of existing indirect reuse situations, and try to mitigate their effects by reducing the discharge of contaminants into water sources and removing the contaminants by treatment, indirect potable reuse is, in my view, no longer an attractive option for water resource planning.

When the revelations concerning the health effects of organics in the Mississippi River led to the passage of the Safe Drinking Water Act in the US in 1974, Vicksburg, Mississippi, abandoned its intake in the Mississippi River and turned to groundwater. While this may not be feasible for many large supplies, we should certainly be cautious about building new intakes that can be expected to be vulnerable to urban and industrial pollution. The US Environmental Protection Agency in 1978 published a volume of guidelines for location of water supply intakes downstream of effluent discharges, with case studies recording the serious reservations concerning this practice.

Indirect potable reuse by recharge of aquifers with wastewater is currently undergoing critical review. Such indirect reuse offers advantages over surface water reuse, because percolation through soil provides greater treatment. More important, the movement of water underground is slow, and observation wells can be used to detect changes in water quality so that measures can be taken in the event contamination is detected. On the other hand, once an aquifer becomes contaminated, it may not be feasible to decontaminate it. The state of California has undertaken long-term investigations into the appropriateness of groundwater recharge for potable reuse. At present, in the US, the jury is still out and the engineered recharge of aquifers for potable purposes is still suspect.

DIRECT POTABLE REUSE
The policy of piping wastewater effluent directly into a water supply system intended for drinking has not won many adherents despite the reportedly successful experience in Windhoek many years ago. I know of no intentional direct potable reuse anywhere in the world outside of Southern Africa. Direct potable reuse suffers all the hazards of indirect potable reuse, while abandoning the ministrations of time and nature which provide some mitigation of pollution in indirect reuse.

To some, direct potable reuse appears attractive and/or necessary, and they have undertaken to conduct research in support of this approach. The problem that such researchers face is that, whereas with historic indirect potable reuse, the water is innocent until proven guilty, with direct potable reuse, the water is guilty until proven innocent. To prove that the treatment of wastewater will render it not harmful to health is virtually impossible, because the quality of the wastewater changes continuously and will change in the future with the development of new chemicals, and also because the

3

standards of acceptability for long-term ingestion are themselves uncertain and can be expected to change as more health-related experience develops. Furthermore, the many attitudinal surveys made in the US, particularly in water-short California, have shown little public acceptance of reuse of wastewaters for drinking, where there is acceptance for non-potable purposes.

Capital facilities in the water sector are expected to serve for decades and, in fact, often serve for a century or longer. The uncertainties and risks associated with potable reuse are difficult to justify, particularly where other options are available. These other options include the development of new high quality sources or, if that is not economically feasible, non-potable reuse which permits existing high quality supplies to be conserved for higher uses such as drinking.

INDIRECT NON-POTABLE REUSE

Good examples of indirect non-potable reuse are the abstractions from the River Trent in England for industrial and agricultural use. The Trent was not considered a suitable source for potable supplies, particularly as other options were available. In fact, the Trent research programme of the Water Resources Board in 1972 introduced the dual water supply system approach, with the non-potable supply drawn from the Trent.

Opportunities for indirect non-potable reuse can be expected to decrease as direct non-potable reuse is adopted because the latter approach preserves ambient water quality and will in most instances be less costly. Accordingly, emphasis in this paper is on direct non-potable reuse.

DIRECT NON-POTABLE REUSE

Most water supplied does not need to be of drinking water quality. Water for agricultural and urban irrigation, fisheries, recreation, industry, cooling, construction and even many household uses does not need to meet drinking water standards.

Accordingly, non-potable reuse may be attractive in communities or areas where:

(a) freshwater supplies of water are limited in quantity or quality, or where
(b) new water supplies of satisfactory quality can only be developed at high cost because of distance or treatment requirements, and where
(c) a single large water user or a class of users can tolerate water of lower grade than required for drinking.

Non-potable reuse may be lower in cost than other options such as new supplies and it involves less health risk than potable reuse. The economy of such non-potable reuse becomes especially attractive where a high degree of wastewater treatment needs to be provided for pollution control. In fact, the treatment requirements for many non-potable reuse applications may be less costly than for discharge to the environment. For example, where phosphorus and/or nitrogen needs to be removed from an effluent before discharge to a body of water that may become eutrophic, such removal is not necessary if the effluent is used for irrigation.

Non-potable reuse is not new. For almost 50 years, the secondary effluent from the Baltimore wastewater treatment plant has been chlorinated and sold to the Bethlehem Steel Corporation. It is delivered through a 2.4 m dia. pipeline, 7.2 km long. Almost 60 years ago, a dual distribution system was introduced into Grand Canyon village. The potable water is obtained from a spring 1 km below in the canyon, while reclaimed effluent is used in the non-potable system for lawn irrigation in the village and toilet flushing in the hotels and cabins.

What is new is the current widespead interest in and adoption of non-potable reuse schemes. As indicated in Table 1, the bulk of the presentations at the Water Reuse Symposium III were devoted to direct non-potable reuse. The only significant commitment to potable reuse was by South Africa and by Denver in its research programme. On the other hand, some 30 direct non-potable reuse projects were described. California alone has some 380 non-potable projects now in operation providing some 800 000 m^3/day from 240 wastewater reclamation plants. Similarly, schemes in the Middle East provide some 400 000 m^3/day for direct non-potable reuse.

In 1980, the US Environmental Protection Agency published 'Guidelines for water reuse', which was intended to 'increase interest in and assist implementation of wastewater reuse for non-potable purposes: irrigation and agriculture, industrial, recreation, and non-potable domestic use'. In 1983, the American Water Works Association published a manual on dual water systems. In reports to the Science and Engineering Research Council in Britain and the National Science Foundation in the US, Dr Arun Deb had demonstrated the economies that can result from dual distribution systems. Current practice affirms that dual systems are, in certain circumstances, economic. (References to direct non-potable water reuse practices are listed in the Bibliography.)

WATER QUALITY STANDARDS
California has adopted a most extensive set of treatment and quality standards for direct non-potable reuse applications and, in general, these are being emulated elsewhere. A summary of these standards is presented in Table 2.

These standards are seen to become more stringent as exposure to the public increases, both in duration of exposure and in numbers of people exposed. Criteria have not been established for industry as their requirements are generally more stringent than the highest public use, and they can provide the additional treatment. Greatest interest is in the use of reclaimed wastewater for urban distribution in a non-

Table 1. Distribution of subject matter at Water Reuse
Symposium III in San Diego, California, 26-31 August 1984

Type of reuse		Number of presentations
Indirect potable reuse		9
Groundwater recharge	- 6	
Surface water discharge	- 3	
Direct potable reuse		10
Republic of South Africa	- 4	
City of Denver, Colorado	- 4	
Non-potable reuse		60
United States	- 50	
Israel	- 3	
Saudi Arabia	- 3	
Developing countries	- 4	
Industrial recycling		14
Land (including wetland) treatment		13
Miscellaneous		47
	Total	153

Table 2. California wastewater reclamation criteria*

Use	Treatment	Coliform, standard: MPN/100 ml
Irrigation of crops		
Fodder fibre and seed crops	Primary	None
Processed produce; surface irrigation	Primary	None
Processed produce; spray irrigation	Secondary disinfected	23
Produce eaten raw; surface irrigation	Secondary disinfected	2.2
Produce eaten raw; spray irrigated	Secondary coagulated filtered disinfected	2.2
Landscape irrigation		
Golf courses, cemeteries, freeways	Secondary disinfected	23
Parks, playgrounds, schoolyards	Secondary coagulated filtered disinfected	2.2
Homes (fixed sprays), golf courses near homes	Secondary disinfected filtered disinfected	2.2
Recreation impoundments		
No public contact	Secondary disinfected	23
Boating and fishing only	Secondary disinfected	2.2
Body contact (bathing)	Secondary coagulated filtered disinfected	2.2

* Adapted from California Administrative Code, Title 22, Div. 4, Environmental Health, 1978.

potable system, as this involves the widest exposure and the most stringent standards of quality. To assure acceptable quality for general public use, it has been found useful in California to establish a turbidity limit of two units for the effluent of the secondary effluent coagulation filtration reclamation plant. Also, monitoring of chlorine residuals throughout the non-potable system should be much the same as for the potable system. In fact, excessive chlorine, which can assure the prevention of aftergrowths in the system, poses no problem of taste or trihalomethane formation as the water is not to be ingested. The requirement of a technology-based standard, secondary effluent coagulated, filtered and disinfected, was imposed to minimize the virus content in the reclaimed water. Studies at the Pomona plant of the Los Angeles County Sanitation Districts demonstrated that coagulation followed directly by filtration and then chlorination, with low turbidity, was satisfactory.

REUSE APPLICATIONS IN A NON-POTABLE DISTRIBUTION SYSTEM
General aspects of water reclamation for the highest non-potable uses in industrialized countries may be of interest:

1. The water reclamation plant is often located to include consideration of reuse. Whereas a treatment plant is generally located on the outskirts of a community for economies in collection and disposal, a reclamation plant treating only a portion of the total wastewater flow may be located near the points of use. In such instances, the wastewater is withdrawn upstream on the sewerage system. The sludge produced in the reclamation plant is put back into the sewer to be treated at the main plant. This is done in the Los Angeles County Sanitation Districts' reclamation plants.
2. Where urban irrigation is the principal use, as contrasted with use in industry, the requirements are seasonal and provision may need to be made for storage. In the Irvine Ranch Water District in California, storage is in open impounding reservoirs which permit the growth of algae. Water withdrawn from these reservoirs for distribution is refiltered and rechlorinated. In St Petersburg, Florida, storage is underground in a saline aquifer, from which it is to be withdrawn when the demand exceeds the production of the reclamation plants. Diurnal storage in both systems is in covered tanks.
3. In some dual systems, water for fire protection continues to be taken from the potable system. In others, such as St Petersburg, both systems are designed to provide water for fire protection, with modifications to the non-potable water hydrants to prevent the general public from taking water from them.
4. The Los Angeles County Sanitation Districts operate several reclamation plants, but the districts do not provide water supply, so they are obliged to sell their reclaimed water wholesale to responsible water purveyors, who then distribute

the non-potable water within their service areas. This does pose a problem in quality control. In England and Wales, with integrated water authorities, this would not be a problem.
5. Most regulations do not permit the use of hose bibs on the non-potable system, requiring fixed nozzle sprays for irrigation. Also, spray irrigation is often confined to later night hours to avoid exposure.
6. Except for the aforementioned Grand Canyon village, household users in the US have been restricted to lawn irrigation. In the Jurong industrial estate in Singapore, a dual system for industrial use was established based on the sedimentation, filtration and chlorination of a secondary effluent. More recently, the non-potable system has been extended to provide toilet flushing for some 25 000 residents in high-rise housing estates in the vicinity.
7. Industrial and power plant evaporative cooling is a widely used application of non-potable reuse. Studies have shown that the fallout from the vapour is not hazardous because of the heavy chlorination used for slime control.
8. Special measures are taken to avoid cross-connections between potable and non-potable systems. These include pressure control valves and other devices, colour coding, different piping materials etc. National standards have not yet been promulgated in the US.
9. Regional management offers attractive opportunities for rational use of water. High quality water for potable purposes can be distributed regionally, with non-potable reuse relieving the demand on these limited resources.

CONCLUSION
Direct water reuse for non-potable purposes can be expected to grow wherever water resources are limited. Initially such applications were limited to arid and semi-arid areas. Now, however, the economies of water resource development indicate that non-potable reuse for a wide diversity of purposes is an option that deserves attention wherever water resources are scarce.

BIBLIOGRAPHY
AMERICAN WATER WORKS ASSOCIATION. Reuse, Seminar proceedings, Denver, Colorado, 1975, June, no. 20109.
AMERICAN WATER WORKS ASSOCIATION. Dual distribution systems, Seminar proceedings, Denver, Colorado, 1976, July, no. 20135.
AMERICAN WATER WORKS ASSOCIATION. Dual water systems, manual no. 24, Denver, Colorado, 1983.
AMERICAN WATER WORKS ASSOCIATION RESEARCH FOUNDATION. Proceedings Water Reuse Symposium II, 3 volumes, Denver, Colorado, 1982.
BRUVOLD, W. H. and CROOK, J. Public attitudes toward community wastewater reclamation and reuse, Office of Water Research and Technology, US Dept of Interior, Washington, DC, 1980.

CALIFORNIA STATE WATER RESOURCES CONTROL BOARD. Report of the consulting panel on health aspects of wastewater reclamation for groundwater recharge, Sacramento, 1976, June.

CALIFORNIA STATE WATER RESOURCES CONTROL BOARD. Irrigation with reclaimed municipal wastewater - a guidance manual, report no. 84-1 wr, Sacramento, 1984, July.

CAMP, DRESSER and McKEE Inc. Guidelines for water reuse (US Environmental Protection Agency), Boston, Massachusetts, 1980, March.

DEB, A. K. Multiple water supply approach for urban management, National Science Foundation, Washington, DC, ENV-76-18499, 1978, Nov.

LOS ANGELES COUNTY SANITATION DISTRICTS. Pomona virus study - final report, California State Water Resources Control Board, Sacramento, 1977.

NATIONAL RESEARCH COUNCIL. Drinking water and health, 5 volumes, Washington, DC, 1977-1983.

NATIONAL RESEARCH COUNCIL. Quality criteria for water reuse, Washington, DC, 1982.

OKUN, D. A. Regionalization of water management, a revolution in England and Wales, Applied Science Publishers Ltd, London, 1977.

US ENVIRONMENTAL PROTECTION AGENCY. Guidance for planning the location of water supply intakes downstream from municipal wastewater treatment facilities, 68-01-4473, 1978, April.

WATER POLLUTION CONTROL FEDERATION. Water reuse, manual of practice SM-3, Washington, DC, 1983.

WATER RESOURCES BOARD. The Trent research programme, dual water supply systems, 1972, Vol. 9.

1 Reuse via rivers for water supply

D. D. YOUNG, BSc, FIWES, FIPHE, MIWPC(Dip), Severn Trent Water Authority, UK

SYNOPSIS   Reuse of sewage effluent via abstraction is as
old as waterborne sewage disposal to rivers used for potable
water supplies. One third of UK supplies are taken from
rivers which contain varying amounts of sewage effluent.
The demand projections made in the early 1970s indicated a
need for the use of supplies from rivers such as the Trent,
entailing new use of intakes with high degrees of reuse.
This stimulated an interest in the health aspects of reuse,
which has outlived these projections. The paper examines
the potential effects of sewage effluent returns upon river
water quality and on the quality of such waters when put
into supply and also discusses the fitness of such supplies
for human consumption.

Historical Background   Sewage and other effluents have been
used for water supply via river abstractions since
waterborne waste disposal became general during the last
century. As one third of the water supplies of UK are drawn
from rivers, most of which contain sewage effluent, one
third of the population depends upon supplies containing
water from effluent. With the use of mains water for
manufacture of beverages and food- stuffs, some ingestion of
reused sewage effluent must be almost universal.

In the early days of this general reuse during the last
century, it led to catastrophic epidemics of water borne
diseases such as Asiatic cholera and typhoid. However, when
the water supply link with these diseases became clear,
solutions were found in the development of alternative
sources such as the great reservoir and aqueduct schemes for
Birmingham, Liverpool and Manchester and other cities, in
the moving of effluent discharges downstream and intakes
upstream as in the case of London, and in the progressive
introduction of water filtration. Ultimately, our present
almost absolute security from infection due to reuse of
sewage effluent was established through the standard
practice of chemical disinfection of water supplies coupled
with increasing sophistication of treatment processes and
management of distribution systems.

The tragic history of the nineteenth century epidemics has left a lagacy of dislike of reuse and a tradition of always using the best available raw water source for public supply with the implication that river intakes should have a minimum of population and industry upstream.

Influence of Demand Projections   The rising demand for water in the early post war years, the growing public objection to new reservoir and groundwater development schemes and the progressive commitment of the more economic sources, led to the passage of the Water Resources Act 1963 with objectives which included "securing the protection and proper use of inland waters and water in underground strata". The Act set up the River Authorities with duties which included the "conserving, redistributing or otherwise augmenting of water resources and securing their best use", and a Water Resources Board to advise river authorities and government on these matters. The provisions for raising charges to finance the work of water resource management set out a number of criteria for determining the amount of charge for an abstraction, including "the way in which the water is to be disposed of".

These influences led to a new school of thought on intake selection in which the discharge of effluents downstream of intakes became seen as "a loss to resources" which had to be made good by the provision of other resources through progessively more expensive schemes. This was a reversal of the tradition of use of the best available raw water sources. Abstraction charging schemes made by some river authorities carried this concept through to the point where "return to resources", in the form of effluent discharge upstream of intakes, was rewarded by an abatement of the licence charge.

The 1963 Act also required surveys of water resources to be carried out at the local and national level by the river authorities and the Water Resources Board respectively. These surveys led in the early 1970s to predictions of rapid and continuing growth of demand for public water supply and of production of effluent which were expected to double by the end of the century and to continue growing thereafter.

These prediction have not stood the test of time as exemplified in Table 1.

Table 1   Public Water Supply Demand Forecasts for the
           Severn and Trent Catchments

DATE OF FORECAST	FORECAST DEMAND FOR 2001 (Ml/d)
1971	3912
1975	2478
1980	2393
1984	2212

Some growth is still anticipated, the current forecast for
2001 being some 15% greater than current demand.

Not only did these over estimates of future growth increase
pressure for use of effluent laden rivers, but they were
also accompanied by predictions of increase in effluent
content of rivers already used for public supply. One of
the more noteworthy proposals for meeting the forecast
demands was the suggested supply of water drawn from the
River Trent at Nottingham to serve the East Midlands.

The Extent of Reuse This is touched on by Packham in a
recent review (1). Some figures assembled in the course of
another unpublished Severn Trent study are set out in
Table 2.

Table 2   Estimates of Effluent Content of Trent and
           Established Water Supply Rivers at Relevant
           Points (Mid 1970s)

	Trent	Thames	Severn	Derwent
Direct industrial discharges excluding cooling water (Ml/d) (A)	510	254	69	17
Total sewage effluent (Ml/d)(B)	1484	789	554	172
Industrial content of sewage effluent (Ml/d) (C)	290	104	24	33
Total effluent content (Ml/d) (A+B)	1994	1043	623	189
Total industrial effluent content (Ml/d) (A+C)	800	358	93	50
Mean river flow (Ml/d)	7500	5720	9230	1700
95% low river flow (Ml/d)	2330	800	2000	500
% effluent content @ mean flow	27%	18%	7%	11%
% " " @ 95% low flow	86%	130%	31%	38%
% industrial effluent @ mean flow	11%	6%	1%	3%
% " " @ 95% low flow	34%	44%	5%	10%

Many speculations may be based on these comparisons but it is clear that the proposition to draw public water supplies from the River Trent near Nottingham did not involve acceptance of any new threshold of sewage effluent reuse, although it did involve a relatively high average proportion of water of industrial origin. The conclusions of the Trent study, described below had undertones of double standards, but it was more difficult to make a conscious decision to introduce a new major high reuse source than to accept continuation of use of established high reuse sources.

The River Derwent scheme cited in the last column of Table 2 was commissioned in 1970 and is situated downstream of Derby and of a large mixed chemical manufacturing concern. Since that time, the intake has been closed two or three times per annum due to pollution incidents as an indirect consequence of reuse. The uncertainties of the quality of this supply, coupled with the availability of plentiful future supplies supported by Carsington Reservoir has led to the replacement of the intake by another from the same river upstream of Derby to be commissioned in 1985.

Wholesomeness and Reuse   The anticipated increased reuse of drinking water was bound to lead to concern as to the wholesomeness of the supplies in question. This related not only to conformity with standards of chemical quality but also to any other possibility of long term injury to consumers due to subsistence upon supplies with a relatively high effluent content. This has broadened into a concern with the long term health implications of the organic content of all waters particularly after modification by disinfection. These issues became prominent in the study of the possibilities of supplies from the Trent at Nottingham.

The study of treatment of Trent water by the Water Research Association has been described by Miller and Short (2) and by others. Their work demonstrated that such water could certainly be treated to meet any contemporary quality standards, albeit at fairly high cost. However, they stated that the Association was not responsible for any decision as to whether such a treated water was "wholesome" or not.

In the report of the Water Resources Board on the Trent study (3) this caveat was strengthened to "this should produce a potable water, though not one which would necessarily be accepted as wholesome by a water undertaker." The Board's conclusions emphasised that it was not their responsibility to decide whether treated water is wholesome but then took that responsibility in part by saying that it was extremely doubtful whether Trent water so treated would be regarded as acceptable for public supply.

Commenting upon the problem of definition of "wholesomeness" which prevented them from coming to a definite conclusion, they urged the Department of the Environment to direct its attention to this problem without delay. In recent years the whole issue of water quality and health including the effects of reuse has been studied in a major programme at Water Research Centre sponsored by the Department.

This programme, and other work, seeks harmful effects on consumers due to residual contaminants associated with reuse present in public water supplies. To the author's knowledge, no such effects have yet been established, although it has not been proved that no such harm can arise either. Such a negative case is probably as difficult to establish in water supply as in almost any other field of human activity.

One of the most interesting studies is that by Beresford (4), who conducted a retrospective study of "the mortality experience from different causes, principally cancer, for 29 boroughs and districts in the London area for a period of six years". Some of these boroughs had been supplied exclusively with well waters, others with mixtures of well and surface waters and others solely with waters from the lower reaches of the Thames and Lee both of which contain a high proportion of reused water.

The study showed no association between high levels of water reuse and cancers of the gastro-intestinal and urinary tracts, when socio-economic differences and variations in population size between the boroughs were taken into account. This study was the precursor of a larger study on a national scale, now in progress and now believed to be approaching its conclusion.

Reuse and Mineral Content The increment of concentration of inorganic contaminants through domestic use is highly variable, being influenced by differing social habits and water consumption. The extent and variability of this effect, based on an American study (5) is indicated by data in Table 3 showing the increase between tap and sewer.

Table 3   Range of Mineral Addition
             Through Domestic Use (mg/l)

	RANGE	LIMITS +	
Calcium	6-109	100(GL)	
Magnesium	4-111	30(GL)	50(MAC)
Ammonium *	2- 48		
Sodium	14-742	20(GL)	150(MAC)
Phosphate (as P_2O_5)	6- 50	400(GL)	5000(MAC)
Chloride	22-1262	25(GL)	
Sulphate	8-191	25(GL)	250(MAC)

* oxidised to nitrate in sewage works and/or river.
GL for Nitrate is 25 with MAC of 50 expressed as
nitrogen.

+ GL = Guide Level   MAC = Maximum admissible
                            concentration

The EC Directive permits derogations from the standards,
where they relate to situations arising from the nature and
structure of the ground and permits delays in compliance in
respect of other factors. The areas of concern relate to
the so-called maximum admissible concentrations, but it has
been suggested that even these might be interpreted as
maximum admissible average concentrations and not maximum
admissable instantaneous concentrations.

Experience and advice given to water authorities in recent
years suggests that magnesium (particularly in association
with sulphate) and sodium content might have implications
for health and the wide range of increments reported
indicates that actual or potential concentrations of these
substances merit study in consideration of existing or
proposed reuse via rivers.

The concentrations of nitrate arising from the oxidation of
ammonia present in sewage are such as to cause difficulty in
meeting the standard in extreme cases of reuse, and this is
discussed below.

A few consumers will detect chloride taste at values above
200mg/l and complaints may become more widespread at levels
of 500mg/l and above, again a case for consideration in the
context of any existing or proposed reuse.

Reuse and Nitrate Content This topic has been dealt with
comprehensively by the report of the Nitrate Sub Committee
of the Standing Technical Advisory Committee on Water

Quality (6). They concluded that "in general, enhanced nitrate concentrations in rivers, with water supply problems, are derived from surface runoff from the land". However, sewage effluent nitrate does make a significant contribution to the overall nitrate balance at low flows in those rivers where effluent reuse is high. Fortunately, the development of the anoxic zone technique in the activated sludge process has made it possible to make a substantial reduction in the nitrate content of effluents, while improving the stability of the process and cutting power consumption.

The concern about nitrate concentration in water supplies relates to methaemoglobinaemia (the "blue baby" syndrome) and suggestions that excessive consumption of nitrates may give rise to certain cancers.

Existing standards are set to provide protection against methaemoglobinaemia with a wide margin of safety but no cases in the UK have been attributed to public supplies.

As pointed out by Packham (1), gastric cancer rates have been decreasing while nitrate levels have been increasing and many of the areas showing the highest rates of gastric cancer have a low level of nitrate in their water supply. No significant evidence of a problem has emerged from the specific studies of water nitrate in relation to cancer mortality.

Foodstuffs, particularly processed meats and other foods preserved by addition of nitrate preservatives, are a major contributor to the consumption of nitrate.

Reuse and Organic Content All drinking waters, particularly those derived from upland and lowland waters, contain significant amounts of organic matter, normally measured as Total Organic Carbon (TOC) content. The composition of this organic material is extraordinarily complex, and this complexity is increased by chemical changes in water treatment processes, particularly during chemical disinfection.

Modern techniques, particularly gas chromatography linked with mass spectrometry (GCMS) make it possible to identify a surprisingly large number of different compounds, albeit at very low concentrations in most cases. GCMS is only effective in identifying substances which can be volatilised under the conditions of gas chromatrography without chemical change. Unfortunately this proviso is not met by some 80% of the organic content of many waters. Some progress is being made in characterising this non volatile material by linking mass spectometry to other chromatographic techniques but it is likely that part of the organic content of surface derived drinking waters will defy analysis for many years to come.

Table 4 Contribution of Effluents to River Water TOC
         and to Chlorination Products

	R. SEVERN AT SHREWSBURY	R. SEVERN AT TEWKESBURY	R. DOVE AT EGGINTON	R. BLYTHE AT BLYTHE BRIDGE
Mean flow (Ml/d)	3300	6800	1100	160
TOC load in river Kg/d	11550	30600	3850	1120
Effluent TOC load Kg/d	290	3980	795	420
% TOC from effluent	3%	13%	21%	83%
95% low flow (Ml/d)	600	1120	250	54
TOC load in river Kg/d	2100	6120	1170	485
Effluent TOC load Kg/d	230	3180	640	335
% TOC from effluent	11%	52%	55%	69%
Trihalomethames * (µg/l) mean	109	83	62	75
max.	139	270	80	500

* measured at outlet from treatment plant

This table indicates that sewage effluent contributes a
relatively small proportion of the organic matter content at
mean river flows, becoming a majority contributor at high
reuse intakes for relatively short periods of low flow. In
terms of the water presented to the consumer, the low flow
peak may only be relevant in cases of supply without
bankside storage.

The most easily identified, and hence widely recognised,
group of products of chlorination of organic matter in water
are the tri-halomethanes (THMs), chloroform, bromoform and
iodoform. THMs do have toxic effects but only at dosages
many orders of magnitude greater than those arising from
drinking water from public supplies. They are also regarded
as indicators of the presence of other unknown chlorinated
organic compounds which are not readily detectable by
current analytical techniques and which may or may not
represent a more credible hazard.

The data in Table 4 suggest that the concentration of THMs
in supply is influenced as much by naturally occuring as by
sewage effluent derived organic matter. The peak values
appear to be associated with plankton growth and the overall
level may be influenced much more strongly by prechlorination
practices than by effluent derived organic matter.

Direct health aspects aside, the organic content of surface
derived supplies can cause severe nuisance to consumers.
It is the basis for the growth of micro-organisms within the
distribution system which may give rise to musty odour and
taste and which in turn provide the basis for ecological
communities developing within the system which include
asellus and other harmless but offensive organisms which may
appear at the tap.

The reaction of this organic matter with chlorine in long
distribution systems can consume the residual chlorine put
into the supply and make it difficult to ensure a
satisfactory bacteriological quality at the tap. However,
the data in Table 4 suggest that the contribution to these
problems from effluent reuse is limited.

Reuse and Industry   Table 3 indicated the extent to which
industrial effluents, discharged to the sewers or direct to
stream, may contribute to reuse. This is a route by which
harmful substances might pass from industrial premises into
the water supply which has been studied by the Thames Water
Authority (7) The aim of their approach was to predict the
presence, fate and significance of organic pollutants from
industry which could arise at concentrations down to 0.1μg/l
at the intake. They recognised that certain substances such
as polycyclic aromatic hydrocarbons and dioxins might
require lower criteria than 0.1μg/l. An inventory was then
drawn up, in collaboration with industrialists known to be
making discharges of, or using chemicals which could
conceivably arrive at intakes at concentrations above the
criterion of 0.1μg/l. The available information on
toxicological, mutagenicity and other physiological effects
was considered alongside the biodegradability and potential
for removal of the substance by physico-chemical processes.
In the case of the river Thames some 3,000 such chemicals
were found to require assessment. It was found that the
vast majority of them were likely to be innocuous at the
levels predicted, giving considerable peace of mind if not
absolute proof of safety.

Other authorities have watched the Thames initiative with
interest but have not generally committed the resources
required for this approach.

The systematic analytical examination of river waters has
also occasionally disclosed the presence of contaminants of
industrial origin such as dieldrin and chlorinated phosphate
esters. The levels observed have been well below those
thought to pose any hazard to the consumer, but caution has
led to firm action by the water authority to secure
reduction or elimination of these substances.

Aesthetic Aspects Most people, given a choice, would prefer a supply which does not contain water consumed and excreted by somebody else. However, in a typical high reuse catchment, little more than 0.1% of the mean flow may arise from human excretion.

The relevance of this factor is exemplified by the suggestion made some years ago that the excretion of female hormones from the use of contraceptive pills would have undesirable effects in men in communities served by sources downstream. Taking the worst case assumptions of use by 25% of the whole population, quantitative excretion of the hormones concerned and quantitative passage through sewage treatment plant, river, water treatment plant and distribution system, even 100% reuse could only lead to an inadvertent dose of about 0.03% of the effective dose used. Bearing in mind that these hormones and their derivatives are also present in excretions from female non users of the pill and, in smaller amounts, from men, it was reasonable to conclude that any additional effect from the pill was bound to be small.

Similar measuring calculations would apply to other substances excreted by the population, but the idea of reuse is likely to remain unattractive to the public.

Pollution Hazards Occurrences of interruption of supply or harmful contamination of supplies due to effluent discharges are remarkably rare bearing in mind that one third of UK supplies are drawn from lowland rivers. The three most famous cases in the last decade have involved a direct discharge to stream from an agricultural chemical plant, the contamination of a reservoir due to pyrolosis products from burning tyres travelling through underground strata and the spillage of phenol to a river respectively. None of these involved discharge via a sewage works.

Intake closures due to pollution are more common and have usually been due to transient events, such as road accidents, fires and spillages of oil and other liquids. These are not directly associated with reuse, but they are associated with it in being due to the presence of population and industry. Thus the extent of reuse can be a useful rough indicator of the susceptibility of a source to accidental pollution. Effective pollution control activity does much to reduce these hazards, but it can never eliminate them and they are an essential factor in source selection and protection and in provision of treatment.

Summary and Conclusion The present view of the hazards, if any, related to reuse via rivers has been summed up by Packham (1) in the following words, "It is impossible at this stage of the research to draw definite conclusions concerning any risks to health arising from waste water recycling. If there are any such risks then they seem likely to be small but the lack of information on the large non volatile group of organic compounds is an important deficiency in this work at present".

The experience of the river Trent study combined with the long standing use of lowland rivers with a relatively high effluent content, shows that we are generally prepared to accept existing practice of reuse via rivers but reluctant to carry it further in proposals for new sources. This is a tenable position with the prevailing modest projections of growth of demand and our increasing ability to exploit existing resources by conjunctive use, linking of demand areas, and so forth.

Ultimately, this issue will have to be faced in the form of future choices between sources involving high degrees of reuse on one hand and alternatives involving higher cost and potential harm to the environment by flooding large tracts of land or diminishing the base flow of streams by groundwater abstraction on the other.

The urgency of this question will return if, and when, demand projections turn upwards again. The answers will take years of work to obtain and it is essential for these investigations to continue.

Acknowledgement and Disclaimer The author is grateful to his Authority for permission to present this paper but emphasises that the views expressed are purely his own.

References

(1) Packham R.F. "Water quality and health"
      Water Bulletin (May 1984) No.107,
      Water Authorities Association

(2) Miller D.G. and Short C.S.
      "Treatability of River Trent Water"
      IWE Symposium on advanced techniques of river basin
      management: the Trent model research programme" 1972

(3) Water Resources Boad "The Trent Research Programme -
                    Vol. 1" HMSO 1973

(4) Beresford S.A.A. "The Relationship Between Water Quality
      and Health in the London Area" - International Journal
      of Epidimiology (1981), Vol. 10, No.2.

(5) Evans R.L.
      "Addition of Common Ions from Domestic Use of Water"
      American Water Works Association (1965), Vol.60, No.3.

(6) Standing Technical Advisory Committee on Water Quality -
    Fourth Biennial Report (February 1981 - March 1983)
    Appendix A   HMSO 1984

(7) Richardson M.L. and Bowron J.M. (Thames Water Authority)
    "Catchment Quality Control"
    Notes on Water Research No.32,
    Water Research Centre, January 1983.

2 The urgent need for the reuse of sewage effluent in developing countries

B. Z. DIAMANT, FICE, FIWES, FIPHE, FRSH, Professor of Environmental
Health Engineering, Department of Water Resources and Environmental
Engineering, Ahmadu Bello University, Zaria, Nigeria

SYNOPSIS. Soaring urbanization trends have aggravated the
wastewater disposal problems in many developing countries cities
urging them to embark on central sewerage systems for solution.
Almost all these systems, that were mainly designed by foreign
consultants, incorporated the solution for disposal, by means of
dilution in natural bodies of water. Since most developing
countries people drink water from raw unsafe surface sources,
this disposal method poses a dangerous health hazard, even if
full treament is applied to the wastewater prior to disposal.
The appropriate disposal replacement method is land application by means of irrigated agriculture, that has to be practiced
in line with certified and approved public health precautions.

INTRODUCTION

1. Steadily growing urbanization trends have been noticed
in many developing countries in recent years. This social process, that had started to affect the developed countries during
the Industrial **Revolution in the middle** of the 18-th Century,
has been emerging in the developing countries only after obtaining independence during the second half of our century. The
urban population in Third World countries is, therefore, low in
comparison with the industrialized countries, but it is growing
steadily and fast. In 1980 some 31 % of developing countries
people lived in urban communities, as against 77 % in the developed countries (ref. 1). The World Bank has carried out a
population survey in the developing countries (ref. 2). The findings of the survey revealed huge urbanization growths towards
the end of the millenium (see Fig. 1). The largest growth has
been noticed in Latin-America, where not less than 75 % of the
population were expected to live in urban communities by the
year 2000. This issue poses a major problem for the Third World.

2. Meanwhile, the cities in the developing countries, in par-
ticulat the Capitals, were cought unprepared by the surging
urbanization trend. Most of them lacked the basic public facilities, like safe water supply and proper waste disposal services, not mentioning primary needs like housing and education.
Many expanding urban areas have come under public pressure and
have embarked, in recent years, on modern, costly, large-scale
water supply and waste disposal programmes, with the assistance

Reuse of sewage effluent. Thomas Telford Ltd, London, 1984

23

Percent urban
population

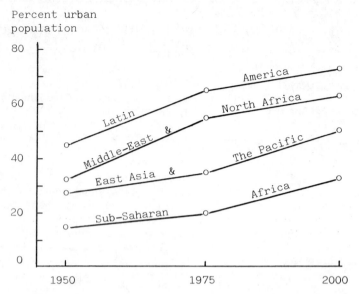

Fig. 1   Urbanization trends in developing countries

of national - and quite often also international - resources.
due to the chronic shortage in experienced local professional
engineering manpower in most developing countries, almost all
planning and design works for these large-scale water supply
and wastewater disposal schemes, have been performed by foreign
consultants, mainly from the developed countries. Unfortunately,
many of these consultants, though well experienced in their
skills, often lacked ample knowledge with regard to environmental health conditions that were existing in their countries of
assignment, which were entirely different from those existing in
the developed countries. As a result, engineering solutions that
were quite adequate for the latter, did not fit at all when
applied in the former countries.

   3.   One of the most serious mistakes performed accordingly,
refers to the disposal of sewage effluent. The conventional
effluent disposal method in the developed countries has always
been performed by means of dilution in natural bodies of water.
The dilution method of disposal is based on the re-oxygenation
and self-purification natural processes that help purifying the
diluting bodies of water (ref. 3). The main concern of the designer applying this method is directed normally to the maintenance of aerobic conditions in the receiving water, so as to avoid
sanitary nuisances, mainly obnoxious smells, and to preserve the
aquatic life in water. The suitability of the receiving water
for human consumption did not pose an environmental problem, because all people living in the developed countries, including
those living in the rural areas, have had access to safe drinking water. This is the reason why wastewater disposal by means

of dilution in natural water courses, can be accepted and tolerated in the developed countries. Engineering consultants from these countries that undertook the planning and design of wastewater treatment and disposal schemes in developing countries, have wrongly assumed that technological solutions that were functioning satisfactorily in their home countries, could be applied as well in the developing countries. Numerous central sewerage systems have been designed in recent year for developing countries cities and almost all of them have adopted the dangerous dilution disposal method. Fortunately enough, most of these schemes are still in the design stage, so that it is yet not too late to remove this wrong method from the plans and replace it with a more suitable and appropriate solution. Before discussing the appropriate alternative, it might be worthwhile to review the drinking water supply conditions in the developing countries, inorder to achieve a better assessment of the sever environmental repercussions involved in practicing the conventional method of wastewater disposal by means of dilution in natural bodies of water.

WATER SUPPLY IN THE DEVELOPING WORLD

1. The rarity of safe drinking water in the developing countries is a well-known fact, but the alarming extent of this rarity has been fully exposed for the first time by the findings of the Global Survey on Water Supply and Wastewater Disposal in Developing Countries, carried out by the World Health Organisation during the previous decade. An Interim Report, covering the first half of the decade (1970-75), was released in 1976 and revealed surprising facts and figures (ref. 4). The Report informed on vast differences between the urban and the rural sectors, in the availability of the two major services that were assessed : water supply and waste disposal.While 77 % of third world urban people had access to safe drinking water in 1975, only 22 % of the rural people had this privilege. This was a very serious finding, in view of the fact that 7 out of every 10 developing countries people, were living in rural areas. This inequality gap was found to be by far wider with waste disposal facilities, when only 12 % of the rural people had proper waste disposal facilities in 1975, against 72 % of the urban people provided with this vital environmental service. The two major facilities are ofcourse directly interrelated when the quality of a water supply largely depends on a local proper waste disposal.

2. Based on these findings, the Interim Report contained also predicted improvements in the severe situation towards the end of the decade in 1980. Accordingly, the availability of safe drinking water in the rural areas , was to cover 36 % of the people instead of 22 % in 1975. Adequate waste disposal was expected to cover 22 % instead of 12 % in 1975. These great expectations, still left 2 out of 3 rural people without safe drinking water, and 3 out of 4 people without proper waste disposal means by 1980. Ironically enough, even these humble expectations did not come true. A later survey carried out in

1980 (ref. 5), revealed very poor improvements indeed in water and waste disposal, in particular in the rural sector. Safe rural drinking water supply population coverage has been increased only by 7 % (from 22 % in 1975 to 29 % in 1980), and the coverage for waste disposal - by only 1 % (from 12 % in 1975 to 13 % in 1980). The findings of the 1980 survey in the urban areas clearly reflected the impact of the urbanization trend, which has caused a deterioration in the environmental conditions. The urban safe drinking water coverage has dropped from 77 % in 1975 to 75 % in 1980, and in waste disposal the drop has been even worse, from 72 % in 1975 to 53 % in 1980.

3. Nevertheless, the urban population in the developing countries is still by far better off than the rural one. It should be noted that the urban population growth in the developing world has often exceeded many times the average natural growth. In Bangkok for example, the population has quadrupled in the course of the last 2 decades, from 1.25 million in 1960, to over 5 million in 1980 (ref. 6). The above-mentioned poor environmental conditions in the developing countries have had very grave repercussions. Out of the 13.6 million children (up to age 5) that died in the world in 1980, 13.1 million were from the developing countries, where most deaths were related to water-borne diseases and "could have been prevented" (ref. 7).

4. The W.H.O. Global Survey had a strong impact on the U.N. Water Conference, held in Mar-del-Plata, Argentina, in 1977. In its resolutions the Conference requested "a commitment of all national Governments to provide all people with water of safe quality and adequate quantity and basic sanitary facilities by the year 1990." The Conference has further recommended to declare the decade 1981-1990 as the International Drinking Water Supply and Sanitation Decade (IDWSSD). In November 1980 the U.N. General Assembly officially declared and inaugurated the IDWSSD. The cost of the programme has been estimated at the huge sum of $ 300,000,000,000 during its 10-year duration, making it one of the most ambitious international undertakings. The magnitude symbolises the importance of - and the top priority assigned to - the issue of available safe drinking water and proper sanitation in the developing world.

THE CONTAMINATION IMPACT

1. Domestic wastewater is classified among the most dangerous contaminants in the environment, because of its human-waste component. Various purification rates are applied to domestic wastewater inorder to eliminate, or at least reduce, its contamination potential. The purification process is normaly carried out in 2 stages. The primary stage is quite simple and is based on physical reactions, such as screening and sedimentation.This stage can remove about 40 % of the original contamination in wastewater and reduce its biochemical oxygen demand (B.O.D.) by the same rate. When this reduction in strength is not sufficient for safe disposal, a more costly and complicated stage has to follow, known as the secondary treatment.

This stage can remove additional 50 % of the strength, bringing about an overall reduction of 90 % in a 2-stage treatment plant. In certain circumstances, when large quantities of industrial wastes are mixed with the domestic wastewater (which is quite common in large urban communities), a tertiary treatment stage is added to the treatment. This stage involves more costly processes based on chemical precipitation and reaction.

2. Since the most dangerous component in domestic wastewater is the human-waste, its contamination is assessed according to the concentration of intestinal bacteria present in the contaminated water. The bacteria which is used for the assessment is an intestinal bacteria of the coliform group known as B. coli. The presence of B. coli in water normally indicates faecal contamination with human waste, which can cause dangerous water-borne diseases such as typhoid fever and cholera, among people drinking this water. The standard assessment is expressed in the concentration of B.coli bacteria in 100 ml of the water sample. The upper permissible limit for this concentration is considered to be as follows : 10/100 ml in rural areas and 2/100 ml in urban areas where the density is higher and the spread of disease is more acute (ref. 8).

3. A full-scale wastewater treatment plant incorporating the primary and secondary stages, can reach a 90 % purification rate and in exceptional cases, even up to 95 %. Nevertheless, inorder to fully assess the contamination impact of the dilution disposal method on natural water courses serving as receipients, let us a^{assume} theoretical wastewater treatment plant that is capable of performing a fantastic 99.99 % purification rate. The effluent of this plant will contain only 0.01 % of the original contamination of the raw domestic wastewater. The impact of this small remaining rate of contamination can be assessed as follows (ref. 9) : A 100 ml sample of domestic wastewater contains approximately 2,000,000,000 B. coli bacteria. The above effluent will hence contain -

$$2,000,000,000 \times \frac{0.01}{100} = 200,000 \: / \: 100 \; ml \qquad (1)$$

when this effluent is diluted in a large river with an excellent dilution capacity of 1:10 , the bacteria concentration in the river will be -

$$200,000 \times \frac{1}{10} = 20,000 \: / \: 100 \; ml \qquad (2)$$

Assuming further that half of this contamination will be removed by the self-purification process which takes place in flowing water courses (ref. 3), then the final concentration of B. coli bacteria in the river will still be -

$$20,000 \times \tfrac{1}{2} = 10,000 \: / \: 100 \; ml \qquad (3)$$

This assessment means that the effluent of a theoretical treatment plant providing a fantastic purification rate of 99.99 %, that is diluted in very favourable dilution rates (1:10), followed by a generous self-purification process (50 %), still contaminates the receiving water at a rate 1,000 times higher

than the upper permissible limit of 10 <u>B. coli</u> bacteria in
100 ml of water. Though the contamination of the river is accomplished in the urban area, the infected water will reach
very soon, with the river flow, the rural areas, where people
usually use the raw water for domestic needs. The goal of the
IDWSSD has been the provision of all people of the world with
safe drinking water and proper sanitation facilities by 1990.
Unitl this goal has been achieved, disposal of wastewater in
natural water courses in the developing countires, must be
strictly banned by all means, no matter what rate of treatment
and purification has been applied to the wastewater.

4.   The alternative disposal method to dilution, which is most
suitable for developing countries, is land application by means
of agricultural irrigation. This disposal method has some important advantages on the conventional dilution method. The most
importnat of these is, ofcourse, the prevention of stream pollution. Other advantages refer to the use of the wastewater for
the raising of crop instead of wasting it in the river, and to
the use of the organic matter present in the wastewater as crop
fertilizer. Most wastewaters used for irrigation, reach a satisfactory quality rate for this method of disposal, after primary treatment, saving hence the costly secondary treatment processes. The most suitable treatment for wastewater used for irr-
gation purposes in developing countries, is performed by means
of waste stabilization ponds. However, due to the dangerous nature of this irrigation water, certain precautions must be taken
in practicing this disposal method, inorder to protect the health of the people employed in operation, as well as the health
of the consumers of the raised crops, and of the people living
in the vicinity of the irrigation system.

WASTEWATER IRRIGATION

1.   Irrigated agriculture has been practiced by man since
ancient times. Traces of irrigation systems discovered by arch-
eologists in Egypt, have been estimated to be 5,000 years old.
Irrigation has been defined as "the application of water to soil
for the purpose of supplying the moisture essential for plant
growth" (ref.10). Unlike freshwater irrigation, wastewater
irrigation, due to its dangerous nature, requires special arrangements and precautions as mentioned above. These refer to the
chemical quality of the wastewater, the types of crops raised,
the method of irrigation and the irrigated land.

2.   <u>Chemical quality</u>.   The chemical content of the irrigating
water has a direct impact on the growth of the plants and on the
cultivated land. Among the most common hazardeous chemicals in
irrigation water is the sodium which affects mainly the cultivated land. Sodium causes flocculation of the soil particles and
encourages, hence, compaction, both of which reduce the poros-
ity and impair the Water/Air ratio in the plant. Most soils
contain calcium and magnesium held in an exchangeable form.
Excess dissolved sodium in the irrigation water tends to replace the 2 chemicals in the soil. The sodium hazard in irrigat-

ion water is expressed in terms of Percent Soluble Sodium
(PSS). The PSS represents the ratio, in meq/l, between the sodium cation and the sum of 4 common cations, including sodium:

$$PSS = \frac{100 \times Na^+}{Na^+ + Ca^{++} + Mg^{++} + K^+}$$

The upper permissible limit for the PSS is considered to be
60 %. The significance of the sodium hazard in wastewater used
for irrigation, is due to the fact that the concentration of
sodium in wastewater is usually about 10 % higher than the sodium concentration in the original freshwater that forms the
wastewater. It should be noted that the sodium content in wastewater is not affected or reduced by ordinary wastewater treatment and purificatin processes.

3. Another chemical hazard typical to wastewater irrigation,
is the heavy metals content, which is contributed mainly by
industrial wastewaters collected together with the domestic
wastes in the central sewerage system. Heavy metals like cadmium
nickel, molybdenum etc., are poisonous to plants and through them
to man and animal consuming the infected plants. When dangerous
concentrations of heavy metals are detected in wastewater used
for irrigation, the sources of pollution have to be traced and
the relevant industries responsible for the damage, have to apply pre-treatment stages to their wastewater, prior to disposing
of in the central system, inorder to remove and reduce the high
concentrations of heavy metals. The pre-treatment involves usu-
lly costly chemical precipitation (ref. 11).

4. The crop.   Due to the dangerous human waste component in
domestic wastewater, crops irrigated with this water should be-
long to the non-edible types, such as cotton or timber. However,
when this restriction can not be followed due to agronomical
reasons, then only edible crops that require cooking prior to
consumption should be allowed to be raised, such as egg-plants
and potatoes, but never lettuce, tomatoes or cucumbers. The
selection of the crop will be decided mainly according to agricultural considerations, but due attention must be given to
the main purpose of wastewater irrigation, which is safe disposal. All other involved factors, such as yield and crop quality
are secondary to the main purpose. Therefore, crops that require
large quantities of water, should be preferred on crops that need
less water for growing. This factor, together with the length
of time required for the growth of the plant determine the quantities of wastewater that can be disposed of in a given area of
land. Table 1 contains 10 different kinds of crops, their growing time and water requirements (ref.12). It is quite clear
from Table 1 that rice and sunflowers crops should be preferred on, say, wheat or soybeans whose water requirements are by
far lower for the same area of land.

Table 1.   Water requirements by various crops

Crop	Growing time (days)	Applied Water (cm)	cm water/ 100 days
Rice	98	104	106
Sunflower	110	87	79
Sugar cane	360	237	58
Cotton	200	105	53
Maize	100	44	44
Wheat	88	37	42
Linsead	88	32	36
Soybean	110	37	34

5. <u>Mode of irrigation.</u> The mode of irrigation with waste-water has to be in line with the hazardeous nature of this water. This is the reason why sprinkler must not be used for this type of irrigation, inorder to avoid the spread of conta-minated droplets in the air. Out of the other modes which incl-ude sub-surface irrigation, furrow, strip and basin irrigation applications, the safest will be, ofcourse, the sub-surface mode, but this can not always be applied. Out of the other 3 possibilities,the furrow irrigation should be preferred,since the wastewater applied according to this mode, is confined bet-ween crop beds in the furrows, which reduces the contact of the wastewater with the upper parts of the plant (ref. 3).

6. <u>The Land.</u> The area irrigated with wastewater is usually known as a wastewater-farm (or sewage-farm). The land for the farm is allocated by the local authority in charge of the cent-ral sewerage system. This land has to be designated very clearly in the town-planning scheme for this purpose only, inorder to secure the use of this land. Failing to do so, can result in a regretful return to the dangerous dilution disposal method.The size of the allocated land must consider expansion and future needs of the growing community connected to the system, for a period of 20-25 years. Since the disposal of the wastewater is the main aim of the farm, regular irrigation applications must be carried out even during rainy seasons, though in reduced quantities, while the extra wastewater is stored in operational reservoirs for use in sunny days. For this purpose the ordinary area of land calculated for irrigation, will have to be increa-sed by 25 % (ref.3).

7. The calculated area of land for the farm is determined largely by the type of available soil. Soils can be divided roughly into 5 categories, each bearing a different hydrau-lic loading, which is measured in cubic meters of water per

hectare per day. Table 2 contain the various hydraulic loadings for the 5 types of soil (ref. 12).

Table 2. Maximum hydraulic loadings for various soils

Type of soil	Load (m^3/hectare/day)
Sand	200
Sandy loam	150
Loam	100
Clay loam	50
Clay	30

PUBLIC HEALTH CONSIDERATIONS

1. A wastewater farm must be fenced with signs hanging along
the fence indicating in red letters the dangerous nature of the
place. The piping system carrying wastewater has to painted in
red to distinguish it from freshwater piping. The staff employed
in the farm has to be provided with rubber boots. Hot water
showers have to be built in the farm, to enable the employees
to wash before returning home after work. A proper dinning
place equipped with safe water for drinking and washing , has
to be provided for the employees, as well as semi-annual medical
examinations for early detection of any possible intestinal
and helminthic diseases. The farm itself has to be located in
a distance of at least 1 kilometre from the nearest dwellings,
preferebly in the direction of the prevailing winds away from
the houses, so as to prevent any possible sanitary nuisance in
case of a break-down in the operation of the farm.

2. A wrongly designed irrigation system can lead to a sharp
increase in the rate of vector water-borne diseases in the area,
by providing new breeding places for the disease vectors. The
most common vecor water-borne diseases in the developing countries are malaria, bilharzia and river-blindness, with their respective vectors being the mosquito, the aquatic snail and the
black-fly, all of which require the water media for breeding
and development. It is, therefore, important that vector control
measures should be incorporated and included in the design and
construction of the farm (ref. 13). These include, appropriate
channel slopes inducing velocities exceeding 30 cm/second of the
flowing water, which discourage the breeding of mosquitoes and
snails, clearing of vegetation along channel and reservoir banks
to control mosquito breeding, preventing the creation of small
isolated bodies of stagnant water and the application of intermittent loadings of water, rather than non-stop moderate loadings, which maintain the soil constantly wet and proper for
breeding.

3. Local authorities tend to authorize private bodies to

run and operate public facilities. Normally this is a positive
move since private enterprise is, in most cases, by far more
efficient than salary-paid officials. However, in the case of
a wastewater farm, the local authority should be running and
operating the plant, because too many non-profitable restrictions are involved in the operation of the farm, such as the
limited types of approved crops, excess irrigation in favour of
disposal, health precaution facilities etc. A private contractor, though more efficient, is not likely to observe all these
vital restrictions, and the local authority staff will have to
undertake and assume this duty.

PISCHICULTURE
   1. One of the major nutrition problems in the developing countries is the defficiency in proteins. Fish is an important
source of protein and it can be raised in artificial ponds. These ponds can contain freshwater mixed with wastewater, where the
organic matter available in the latter encourages the growth of
plankton which serves as source of food for fish (ref. 14). This
method of fish raising is highly developed in certain countries
like India and Israel. Experience has shown that yield of fish
ponds mixed with wastewater is by far higher than the yield obtained from ponds containing only fresh water. Furthermore, the
mixed ponds require less added food than the freshwater ponds.
The wastewater irrigation system can, hence, incorporate fish
raising as well, in a way that the treated wastewater is first
directed to the fish ponds and from there to the irrigation area.
The fish ponds can also serve as the operational reservoir where
the wastewater can be stored during rainy periods, as mentioned
above. As far as health precuations are concerned, the fish can
be considered as vegetables that require cooking prior to consumption, which are allowed to be irrigated and raised with wastewater.

   Sewage reuse in the form of wastewater irrigation serves
as an important public-health prevention method in the developing countries, where most people are exposed to risk of consuming raw water from surface contaminated sources. The recent
trend in developing countries cities, to embark on central sewerage systems, with conventional disposal methods, highly increases this risk. The local authorities in charge of the systems,
and, in particular, the foreign consultants that were asked to
undertake the design and constrution of the schemes, must be
aware of this risk and prefer the safe land application disposal
method on the dangerous conventional dilution method.

REFERENCES
1. W.H.O., Global strategy for health for all by the year 2000.
Wld Hlth Org., Geneva, 1981.

2. WORLD BANK, World Development Report. World Bank, Washington, 1979.

3. FAIR, G. M., GEYER, J. C. AND OKUN, S. A., Water and waste water engineering. Wiley, New-York, 1968.

4. W.H.O., Community water supply and wastewater disposal. WHO Chronicle, Vol.30(8), 1976. 329-334.

5. WORLD WATER, D-Day for the Water Decade. World Water, Liverpool, 1981.

6. TREEN, J., Thailand's royal pageant. Newsweek April 1982.

7. W.H.O., 2-nd consultative meeting on the International Drinking Water Supply and Sanitation Decade. WHO/CWS/80.2. Wld Hlth Org., Geneva, 1981.

8. W.H.O., International standards for drinking water. 3rd Ed. Wld Hlth Org., Geneva, 1971

9. DIAMANT, B. Z., The role of environmental engineering in the preventive control of water-borne diseases in developing countries. The Royal Society of Health, Vol.99(3), 1979. 120-126.

10. ISRAELSEN, O. W., and HANSEN, V. E., Irrigation principles and practices. Wiley, New-York, 1967.

11. DIAMANT, B. Z., The control of the heavy metals health hazard in the reclamation of wastewater sludge as agricultural fertilizer. The Royal Society of Health, Vol.101(4), 1981. 127-131.

12. CENTRAL PUBLIC HEALTH & ENVIRONMENTAL ENGINEERING ORGAN- ISATION, Manual of sewerage and sewage treatment. 1st Ed. Ministry of Works and Housing, New-Delhi, 1980.

13. DIAMANT, B. Z., Appropriate technology in the design and construction of irrigation systems in hot climates. Proceedings, Intern. Symposium on Appropriate Technology in Civil Engineering, held by the Institution of Civil Engineering in London on 14-16/4/1980. Paper No. 39.

14. MARA, D. D., Sewage treatment in hot climates. Wiley, Chichester, 1978.

---

3   Reuse of sewage effluent in Australia

A. G. STROM, BCE(Melbourne), FICE, FIEAust, MASCE, MIWES, Gutteridge Haskins and Davey Pty Ltd, Consulting Engineers, Melbourne, Australia

SYNOPSIS. The extent of the reuse of sewage effluent in Australia is discussed and examples given. Whilst widespread in the drier parts, there is little major reuse in the large cities and opportunities remain.

INTRODUCTION

1. In the period 1976 to 1983, the writer directed several studies for the Federal and Victorian Governments into the resource aspects of sewage in Australia, particularly the reuse of sewage effluent (ref. 1, 2, 3 & 4). The following is largely drawn from these.

REUSE IN AUSTRALIA GENERALLY

2. Use of water reclaimed from sewage at present amounts to only about 11% of the total annual sewage flow in Australia. This figure includes the land treatment at Melbourne's Werribee sewage farm, which perhaps strictly should be excluded, as it involves the application of raw sewage to land in summer so as to achieve treatment rather than the reuse of effluent. If excluded, the percentage drops to about 5%. The usage position by States (for locations see Fig. 1) is set out in Table 1.

Table 1. Use of sewage effluent in Australia (1982)

State	Annual Reuse	Total Annual Sewage Flow	Sewage flow reused (%)
	(GL/a)	(GL/a)	
New South Wales	14	550	2.5
Victoria	113	350	32.3
	(23)*		(6.6)*
Queensland	5	200	2.5
Western Australia	8	60	13.3
South Australia	6.8	100	6.8
Northern Territory	2**	6	33.3
Tasmania	Nil	30	0
Total	149	1296	11.4% (4.6%)

* Excluding Werribee sewage farm
** Including projects in course of construction

Reuse of sewage effluent. Thomas Telford Ltd, London, 1984

35

3.   The total annual flow figures are based on records where they are available, otherwise on a 250 L/c/d basis and obviously they are very rough. Even rougher are the reuse figures, as annual records are not kept in most cases and so they have been built up from what information is reasonably available.

Figure 1. Australia

4.   When the limitations of Australia's water resources are considered, perhaps greater reuse could have been anticipated. After all, it is a completely reliable water source, usually close to the population which could use it.

5.   The reasons why reuse is not greater appear to be:
. most of our population lives in seaboard cities, where water supply has not posed major difficulties and the costs have remained at reasonably low levels,

. Sydney-Wollongong-Newcastle, our greatest centre of population and hence of sewage, has shown little interest in reuse in view of its well-watered climate and proximity to the ocean for sewage disposal. (Somewhat similar reasons apply northwards along the seaboard),

. lack of incentive because of present/past water supply subsidies,

- lack of clarity as to health and environmental effects and firm guidelines as to use,

- lack of planning for reuse by the major water and sewerage authorities,

- lack of interest and acceptance by users and public.

6. In the drier States of Western Australia and South Australia, reuse has been much more readily adopted. In the Northern Territory almost all Darwin's dry season sewage effluent is presently planned for reuse, as is all effluent from Alice Springs.

TYPES OF REUSE IN AUSTRALIA
   7. General. These include:

- irrigation for agriculture: pastures; vine growing; vegetables,

- irrigation for silviculture,

- irrigation for landscape: golf courses; parks and gardens; racecourses; mine revegetation,

- municipal: recreational lakes; roadmaking; firefighting; saleyard washing down; domestic garden watering (two-pipe system),

- industrial: coal-washing; dust suppression; process,

- water resource conservation.

8. Types of present reuse State by State are set out in Table 2.

Table 2. Incidents of reuse in Australia - (1982)

	NSW	VIC	QLD	WA	SA	NT
Effluent Reuse						
. Pasture	24	43	20	–	2	
. Vine Growing		1			1	
. Vegetables	1	1			1	
. Trees	2	6	1	1	1	1
. Sportsgrounds (including golf courses)	32	8	30	35	16	2
. Parks	5	2		4	4	2
. Mine Waste Revegetation	1					
. House Garden (two-pipe system)	1				2?	
. Industrial	2			1		

No reuse is known as yet in Tasmania.

9.   Irrigation for agriculture. Most of the incidents of reuse for pasture growing are for disposal rather than planned resource use. Usually there is no charge to the farmer for the water, the prime concern of the authority is to get rid of it with minimum cost and environmental effects.   However, undoubtedly greater numbers of sheep and cattle can be raised in this way and thus the resource is not wasted.

10.   The bogey of "beef measles" (Taenia Saginata) resulting from cattle grazing on effluent-irrigated pastures appears to be a Victorian phenomenon, raised some 50 years ago by farmers hostile to the Melbourne sewerage authority's successful cattle-raising venture on the Werribee farm. For six months of the year about half the raw sewage (240 ML/d) is applied in a cyclical fashion to 8 ha paddocks, which both filter the sewage and grow grass.   Some 15 000 head of cattle and up to 50 000 sheep are maintained, providing an operating profit in good years.

11.   An example of reuse for milk production is that of the Copranapra Pastoral Company, SA, where some 600 cows graze on principally lucerne on 200 ha which is irrigated by pumping secondary effluent from the adjoining outfall of Adelaide's Bolivar Sewage treatment Plant.   Some 280 ha of wine grapes are irrigated with effluent from this same source.   There is a similar wine grape irrigation scheme at Ararat in Victoria.

12.   A comprehensive experimental programme (ref. 5) has recently been completed at Frankston, Victoria, into the watering of vegetables and turf with reclaimed water from Melbourne's 240 ML/d activated sludge plant, the South Eastern Purification Plant (SEPP). Vegetable crops tested included lettuce, carrots, cabbage, celery, spinach and tomatoes.   Irrigation was carried out using overhead sprays. Monitoring of samples from the applied water, the vegetables and the soil was undertaken for bacteriological loading, viral content and heavy metals. Crops were also grown using fresh water and applied fertiliser by conventional market gardening methods, as a control for the experiments.

13.   The work has indicated a good growth response, reflecting the fertiliser content of the water and the better application of fertiliser than by conventional market gardening methods. The bacteriological testing showed the levels of bacteria on the vegetables to be not significantly different to those on vegetables commonly available from normal commercial outlets. The reclaimed water was generally stored for some days before use, and the viral studies showed that, whilst the reclaimed water direct from the outfall channel contained considerable numbers of infectious units (even though it was chlorinated), these reduced to minimal proportions after storage and for the virus experiments, it was necessary to innoculate the reclaimed water with virus before application. After spraying with water containing significant numbers of virus, most were found to die off within 48 hours but small numbers persisted for much longer. The studies indicated that, provided secondary

effluent is held in storage for an adequate period, vegetables spray-irrigated by the effluent should offer negligible risk of viral contamination.

14.   The heavy metal studies also showed that there would be no increased health risk to the community from heavy metal contamination arising from the irrigation of vegetables with reclaimed water. The only significant finding was that of cadmium increase in the control, due to cadmium present in the artificial fertiliser applied.

15.   Irrigation for silviculture. Although incidents are few so far, the irrigation of trees with effluent has strong appeal in many areas.   It permits a variable effluent quality, there is no fear of health problems and in most cases a useful and perhaps profitable crop can be achieved.   Several experimental plantations have been established around Victoria and one at Darwin in the Northern Territory, and there are commercial tree-growing ventures at Alice Springs in the Northern Territory, Wangaratta in Victoria and Dungog in NSW.

16.   The Victorian experimental plantations included a total of 14 species from the genera Eucalyptus, Casuarina, Pinus and Populus planted on a wide range of soil types and irrigated with treated municipal wastewater (winery wastewater at Merbein).   Irrigation applications were metered and recorded and tree growth in terms of height and girth were measured.

17.   The preliminary findings, based on a maximum of four years observations were:

.   the most successful Eucalypt species were Flooded Gum (E. Grandis), Sydney Blue Gum (E. Saligna), River Red Gum (E. Camaldulensis) and Blue Gum (E. Globulus);

.   rapid growth was achieved, as exemplified by four year old Flooded Gum Gum attaining a mean dominant height of 14 m;

.   Pinus Radiata and two clones of Poplar were also found to perform well, as did River Sheoak (Casuarina Cunninghamiana).

The study did not include an economic assessment of irrigated silviculture.

18.   In the commercial tree growing venture at Wangaratta some 30 ha of poplars are spray-irrigated by the local sewerage authority with effluent from lagoons of 20 day retention.   Some 70 mm of effluent is applied each week from October through to April, thereby avoiding any discharge then to the creek system.   The soil overlies well-drained sands.   Although principally an effluent disposal measure, it is intended that the trees be sold to a match manufacturer.   The trees now require careful tending and maintenance to arrive at a satisfactory crop.

19.   At Alice Springs, 50 ha of river red gums have been progressively established over four years and are flood-irrigated with effluent from the primary lagoon of the treatment plant. As well as developing a valuable firewood crop, the scheme partly eliminates discharge from the lagoon system to a swamp which was presenting a health problem. The trees have thrived.

20.   Irrigation for landscape. Irrigation of landscape must be seen as one of the more promising future uses of reclaimed water in Australia. Its value is seen in several ways: by enabling conservation of good quality town supply water, in some cases by postponing or eliminating altogether the need for major augmentation, and in others by increasing the amenity of outback towns where water is scarce and simply could not be used for recreation.

21.   There are now about 80 golf courses irrigated with effluent in Australia and several race courses and sportsground complexes. One of the oldest and most noteworthy is the irrigation scheme based on Adelaide's Glenelg activated sludge treatment plant located about 10 kilometres south-west of the City of Adelaide. Treated sewage has been used since 1953 to irrigate the works itself, and since 1958 the use has been extended by distribution schemes capable of supplying up to 40 ML/d or 90% of the dry weather flow from the works. These supply the following areas:

> three 18 hole and one 9 hole golf courses, and a golf driving range; 2 caravan parks; a bowling green; tennis courts and numerous sporting fields; establishment of grasses for sand dune stabilisation; the surrounds of Adelaide airport; public parks along the Glenelg foreshore and the Patatawalonga River.

22.   Almost half of the allocation, however, is for watering the surrounds of Adelaide Airport. Since no internal reticulation has been installed by the aviation authorities, currently the maximum dry weather usage is probably less than 25 ML/d.

23.   The landscape watering of Broken Hill in New South Wales, is extensive and has been established for over 30 years. It is a remarkable example of the saving in major expenditure on water supply by the provision of amenities by the use of about 6 ML/d of reclaimed water for landscape irrigation. The Zinc Corporation reticulates effluent to the parks and gardens around its lease, to gardens around its offices, pony club grounds, the surrounds of bowling greens, lawns and park around an ornamental lake, sports ovals and domestic gardens of staff. Particular success has been met with the drip irrigation of vegetation cover which is being established for dust control on the very large spoil heaps which surround Broken Hill. It appears that effluent is better for this purpose than fresh water supply even with fertilisers added.

24.   There are several other examples of towns in dry inland areas installing fairly extensive pipelines simply to get effluent to water a park or golf course.

25.   Landscape irrigation is not practised in metropolitan Perth, but in country areas the Public Works Department of Western Australia has taken a constructive lead by insisting that towns supplied with water from subsidised water schemes can only use water on their parks or recreation grounds if all other avenues for supply of water, including treated wastewater, have first been explored. Also, a subsidy was made available towards the cost of the reuse. In consequence, commencing with the town of Merridin in 1972, some 35 country towns, representing most of the inland towns that are sewered, use reclaimed water extensively on public or school sportsgrounds. Treatment is usually by a conventional primary plant followed by a secondary lagoon. Effluent flows to Council's storage dam, which may or may not also contain storm runoff, and is then chlorinated and pumped, either directly or through an elevated pond, to spray irrigate municipal and school sportsgrounds. These simple and relatively inexpensive schemes continue to be highly successful. They remain the only reuse schemes to be subsidised by government in Australia.

26.   In Queensland's Gold Coast area, about 10 sportsgrounds are irrigated with effluent, plus public gardens and the Gold Coast racecourse (and a sub-leased turf farm). In all some 42 ha are involved. Some difficulty is experienced in getting sufficient watering hours because of the high occupancy of the sportsgrounds and the low rate of application necessitated by the clayey soils over what were in the main old refuse fills. The total usage amounts to about 2 ML/d out of a total average discharge from all Gold Coast treatment plants of 43 ML/d. The irrigation schemes have been developed partly as a means of effluent disposal as well as a resource use, and with the future completion of a new major ocean outfall there is likely to be less incentive for further development.

27.   In Victoria the Melbourne and Metropolitan Board of Works, after some years planning, is irrigating a metropolitan park of 118 ha with effluent from a regional plant at Keilor. There appear a number of possible landscape irrigation schemes in the state.

28.   An interesting concept is afforded by the scheme adopted by the Townsville Golf Club, which tapped a municipal sewage rising main and treats 1.6 ML/d in an oxidation ditch. Clarified effluent is used to irrigate the golf course and the secondary sludge developed in the plant is returned to the rising main. This has been an economical and successful project for some 8 years and points the way to future fresh water and money-saving ventures for consideration by the larger city authorities.

29.   <u>Industrial Reuse</u>. This is perhaps surprisingly meagre in Australia and must represent a major growth area in the future. Current reuse includes 7 ML/d for dust suppression and coal washing near Newcastle, NSW and, at Kambalda in WA, effluent from a small treatment plant involving lime precipitation and filtration is used for mine process water and dust control.

30.   Water resources conservation. There are no examples of intentional, planned augmentation of streamflow or of underground water in Australia. However, there is a considerable amount of unplanned augmentation of streamflow resulting from the discharge into inland streams of secondary effluent from the treatment plants of the larger towns and cities. This occurs in about 30 cases in the eastern States. It could be a health matter of concern if town supplies are drawn from the river downstream in view of the known persistence of virus and organics.

31.   Perth's hot climate and lush gardens call for high water usage. It is built largely on sand and has a high winter rainfall. Thus much water is taken by muncipalities, industry and householders from the extensive shallow aquifers under the City and there has been little call for effluent reuse and it mostly is discharged to sea. The Perth Metropolitan Water Board has carried out some experiments into the recharge of the groundwater with secondary effluent from the Kwinana plant.

32.   For several years the Board has been conducting a significant experiment in aquifer recharge in the Canning Vale area, using six spreading basins in the Bassendean sand surrounded by monitoring wells (ref. 6). Application rate is 1 ML/d. The experiments, to extend for another five years, are planned to find the maximum recharge rate, to optimise the removal of nutrients, and to observe destruction of virus and bacteria. The effluent is applied cyclically, usually five days flooding followed by nine days drying out. Early results indicated a progressive decline in the ability of the soil to remove phosphorus as the quartz sands are rather too free of clay and recently by adding red mud (waste product of alumina manufacture) to the sands much better removal of bacteria, virus and phosphorus has been achieved. Nitrates were not being sufficiently denitrified to nitrogen gas with passage through the soil, and nitrate concentration in the groundwater was increasing until biological removal was begun in the treatment plant.

33.   In Victoria the possibility was first raised in 1968 of using a substantial amount of reclaimed water from Melbourne's SEPP to recharge a large and slowly failing rural supply aquifer. Subsequent experimental work on recharging SEPP effluent by borehole to an aquifer near to SEPP brought mixed success, but it is thought that the problems involved could be overcome.

34.   A proposal has been made for experimental work on aquifer recharge near the point where the extensive SEPP outfall enters the ocean. By recharging the considerable depths of calcareous sands through surface spreading basins, it was hypothesised that removal of pathogens and harmful organics would be achieved, permitting renovated water to be withdrawn by pumping for use after softening treatment in the local water system which is presently extended.

35.   Health Matters. Use of reclaimed water is controlled in each State by its Health Department, and reuse schemes are carried out in general accordance with a Commonwealth Government publication (ref. 7).   There have been no reports of any health problems in Australia arising from reuse of effluent.

LOOKING TO THE FUTURE

36.   There is much greater scope in Australia for effluent reuse than at present practised both in numbers of incidence of reuse and in volume of effluent reused. Also, the type of reuse can in many cases be upgraded.

37.   In cities, where large volumes of effluent can be available, industrial usage is seen as being the most economic form of reuse. Industrial water requirements fluctuate little with time and minimum wastewater flows vary little also, so that if the industrial requirement can be matched with wastewater output, vitually all can be utilised. In contrast, irrigation requirements are seasonal and, unless large off-season storages are provided, much of the effluent must be discharged to waste.

38.   The economics of industrial reuse depend upon the quality requirements, for example, for use in food industries, the cost of the many necessary treatment processes would have a major cost impact.   Ideal reuse occurs where only low-grade waters are required, such as coal preparation plants, for dust suppression and for quenching. A further use, and one which is relevant in view of the future substantial increase in thermal power stations, is for cooling make-up water; this is a major demand, generally fairly constant through the year. Reclaimed water was considered recently for Brisbane's Swanbank powerstation.

39.   Accepting that major reuse in the Sydney area appears unlikely because of the configuration of the sewerage schemes, remaining areas for consideration of possible large scale industrial reuse are:

. From Melbourne's South Eastern Purification Plant (SEPP)	240	ML/d
. From Melbourne's Werribee Lagoons	250	ML/d
. Newcastle's Industrial areas,	20	ML/d
. Adelaide's Industrial areas	20	ML/d
. From Brisbane's Luggage Point plant	180	ML/d

A concept proposed for Newcastle was the diversion of dry weather flows only from an ocean outfall to an inland plant for full secondary treatment prior to use in Newcastle's extensive industrial complex.

43

40.   Melbourne's SEPP produces a high grade secondary effluent which is currently pumped 56 km into Bass Strait. The present amount of about 80 GL/annum, a supply unaffected by droughts, will increase by the year 2000 to about 130 GL/annum. At present the possibility of its use for agricultural purposes is being considered, but in the long term potable reuse must be a possibility.

41.   The use in Australian cities of reclaimed water for landscape irrigation is also seen as increasing in future, but the extent may depend upon availability of suitable parks, sportsgrounds and golf courses, etc in reasonable proximity to the sources of effluent. Alternatively, special treatment plants or water reclamation plants on the lines of the Townsville Golf Club for example, could be established to serve particular uses.

42.   Another important future use, although on a small scale, is for landscape irrigation in small towns as a means of improving their amenity at low cost or, in some instances, saving on water scheme augmentation. Tree-growth too is seen likely to be popular in many instances. It is estimated that reuse could be adopted with advantage in about a further 80 Australian towns.

REFERENCES
1.   GUTTERIDGE HASKINS & DAVEY. Planning for the use of sewage. Report for Department of Environment, 3 Vols. May 1976.
2.   GUTTERIDGE HASKINS & DAVEY. Strategies towards the use of reclaimed water in Australia. Report for the Reclaimed Water Committee, Ministry of Water Resources & Water Supply, Victoria. August 1977.
3.   GUTTERIDGE HASKINS & DAVEY. Planning for the use of reclaimed water in Victoria. Report for the Reclaimed Water Committee, Feb. 1978.
4.   GUTTERIDGE HASKINS & DAVEY. Water technology reuse and efficiency. Department of Water Resources & Energy, 1983.
5.   SMITH, M A. Retention of bacteria, viruses and heavy metals on crops irrigated with reclaimed water. Australian Water Resources Council, 1982.
6.   METROPOLITAN WATER BOARD (WA). Canning Vale ground-water recharge project - unpublished reports.
7.   AUSTRALIAN WATER RESOURCES COUNCIL & NATIONAL HEALTH & MEDICAL RESEARCH COUNCIL. Guidelines for the reuse of wastewater, 1980.

Discussion on Papers 1–3

MR D. O. LLOYD, Halcrow (Water)
Considerable development has taken place in many parts of the
world (normally on a regional basis) in the use of systems
analysis techniques to develop computer programs that can be
introduced as tools for government departments to assess
overall water demands and needs. Have the authors been
concerned with similar work?

Halcrow (Water) has been responsible for a master water
resources and agricultural development plan using such
techniques in Qatar.

The agricultural sector of Qatar has for many years
exploited the available groundwater in order to satisfy
irrigation water requirements. This resource is now being
depleted at an increasing rate. In contrast, exploitation of
Qatar's energy resources has provided the country with the
possibility of virtually unlimited supplies of desalinated
water. In addition, with the growth of population, increasing
volumes of treated sewage effluent are becoming available and
can potentially be utilized in the agricultural sector. The
optimal allocation of these three major water resources to
satisfy the future demands of the domestic, commercial and
industrial sectors as well as agriculture formed the principal
objectives of the Qatar study.

In order to allow the range of possible courses of action to
be examined both objectively and efficiently, this study was
carried out using systems analysis techniques, which were
implemented on a microprocessor with 48K bytes of storage. A
series of integrated mathematical models was constructed to
allocate water between competing demands within the country
such that sufficient food was ensured for the inhabitants of
Qatar by maximizing domestic agriculture. This was compared
with imports at a minimum economic cost under any assumed
policy. Primarily the ·following components were critical:
(a) a basic water resources component, including forecasts of
     the availability of groundwater, treated sewage effluent
     and desalination capacity;
(b) an agricultural sector component, including farm budgets
     for three basic farm types – vegetable,
     vegetable/orchard/forage, and orchard/forage, together

with their water requirements; and
(c) a regional water allocation component, which would
    distribute water from the various sources to satisfy
    demands in the agricultural, domestic and industrial
    sectors.

Has Mr Young further information on investigative work
associated with organic compounds and their associated risks
to health if present in potable water? What choices would he
make with reference to the penultimate paragraph of the paper?
   I disagree fundamentally with the overall hypothesis of
Paper 2. It is inaccurate to suggest that consultants were
responsible for the design of sewerage systems and sewage
treatment plants that were totally inappropriate for
developing countries. In my opinion consultants have a duty
to serve the needs of the world and not to remake it in their
own image. All responsible firms respond to demands in an
entirely professional way and do not adopt attitudes and
approaches that Professor Diamant has indicated.
   The main problem in providing adequate sanitation systems in
developing countries is primarily a function of the adequacy
or inadequacy of water treatment and supply systems.
   I totally disagree with Professor Diamant's proposal that
the disposal of wastewater into natural courses in developing
countries should be banned by all means no matter what rate of
treatment and purification had been applied to the wastewater.
Such a proposal is both politically naive and economically
impossible. What method of disposal should be adopted for
cities such as São Paulo, Bogota and Dhaka, if Professor
Diamant's ideas are to be accepted?
   With regard to Paper 3, I refer to the fact that Australia
is probably one of the driest continents in the world and also
the smallest user of water reclamation projects from sewage
treatment plants. How much of the lack of interest in the
reuse of sewage effluent can be attributed to the division
between authority responsibilities within the country? (The
comprehensiveness of UK water authorities enables them to
cover all aspects of the problem.)

DR J. D. SWANWICK, Sir M. MacDonald & Partners, Cambridge, UK
A balanced assessment of reuse has been given in Paper 1, but
I am still concerned that, after over a decade of expensive
research, there is still no clear verdict on the health
effects of indirect reuse of sewage in the UK.
   Paper 2 presents a scenario of a hypothetical 99.99%
purification and yet still strongly condemns indirect reuse
via natural watercourses, in terms of removal of bacteria,
regardless of the degree of treatment. There is no mention of
disinfection in this argument. Would the author not accept,
in view of the high level of treatment, that the further
combined use of storage and disinfection would provide a
satisfactory solution to bacterial and viral contamination, at
least to levels commensurate with those already in the

watercourse before discharge? There are other factors such as
high cost of disinfection, unreliability of the supply of
chlorine and inadequate operational expertise in certain
areas. I am well aware of such problems, but the point is
that in these situations the author's assumption of a high
degree of treatment would probably not be realistic in any
case and a different form of disposal or reuse would be more
appropriate.

The author of Paper 3 appeared to take a more relaxed view,
one that is possibly too tolerant. For example, the reference
to Taenia saginata as a 'bogey' is perhaps an understatement
of that hazard. What period was allowed to elapse between
termination of irrigation and resumption of grazing and what
monitoring of animal health was carried out? Turning to the
trials at Frankston, involving spray irrigation of salad crops
and bearing in mind the author's previous comment that one of
the reasons reuse had not been more widely used was the lack
of firm guidelines·, is the author's statement that such
vegetables 'should offer negligible risk of viral
contamination' to be regarded as his firm recommendation for
future practice?

MR R. W. NASH, Wimpey, London, UK
In my experience, if sewage effluent is recycled in a piped
system, someone is going to drink it. Therefore, if it is not
treated to World Health Organization standards for potable
water it should only be used for specific industrial use,
strictly controlled agricultural use or injected below ground.
It should not even be used for road verge or garden watering.
I have seen roadmen and gardeners drinking 20/30 effluent from
a hose-pipe, with what results I hate to imagine.

In the overseas design and construct package business, we
find that our clients expect the contractor to provide the
scheme, outlined by their professional advisors, at a price
far lower than that which would apply if the criteria provided
was to be worked up into a construction design, and we find
that we have to use our ingenuity, experience and optimization
to achieve an appropriate solution.

Some clients believe in the magic wand approach in that
since nature can produce potable water without the hand of man
for nothing why cannot man do it for the same price?

There are, of course, places in the world where, even if
they caught every drop of rain that fell they still could not
satisfy their demand for water. For these places, safe
recycling is essential to supplement the product of their
expensive desalination plants using enormous quantities of
rapidly depleting stocks of fossil fuels. We must therefore
produce recycled water, to safe potable standards at a
reasonable cost, using a minimum of energy and using
operatives who will never understand the highly technical
apparatus that they are using.

It is a pity that the vast amount of research on the subject
of the reuse of sewage effluent has not been co-ordinated,

47

although such gatherings as this symposium might lead to a less fragmented effort in the future. This international problem should be tackled by an international organization funded by the world's governments.

On behalf of contractors, I would like to point out to Professor Diamant that he should not tar all contractors with the same brush. As more contractors are entering the field of designing, constructing and operating water treatment works of all descriptions, they are being watched more critically than the specialists who have carried out such work previously, and the former unprofessional approach of cutting corners to make more profit is found in the long run not to be profitable. There is also now a professional pride in achieving high standards which brings recommendations and more contracts.

Perhaps if the World Health Organization was to engage appropriate contractors to assist with the research, design and development of projects, the provision of a safe and wholesome supply of water would be available to the whole world's population at an earlier date and at an appropriate price.

MR U. ALKA, Department of Civil Engineering, University of Newcastle upon Tyne, UK

I am an engineer with a water board in Nigeria and would like to point out that what Professor Diamant has said about the use of raw river water by the rural and urban fringe population in developing countries is very real. People are more anxious about locating water rather than defining its quality. From my experience and also recent BBC television documentaries on Latin American countries, it is a common thing to see people taking water from city gutters and streams for domestic use and it is also known that many of the pathogens and contaminants present in wastewater are strongly resistant to the most advanced treatment processes.

In most developing countries there is an extreme shortage of manpower with the necessary know-how for the design and operation of water and wastewater treatment processes, and the expertise of the consulting engineer is invariably relied upon. Therefore consultants must accept that the total responsibility of the environmental acceptability of the designs rests on them, and if these designs are environmentally inadequate the blame should rest squarely on their shoulders.

PROFESSOR D. A. OKUN, University of North Carolina, USA

The California reuse standards include both technology-based standards as well as coliform standards. The low turbidity effluents of the treatment system assures destruction of viruses where the use is accompanied by high exposure.

Cross-connection or inadvertent exposure to the reclaimed effluent will not pose a problem because the water will be biologically safe. The organics, including trihalomethanes, in the water are not a problem because they are only a long-

term health hazard. Also, adequate plumbing codes prevent
cross-connections.

DR J. W. RIDGWAY, Water Research Centre, Medmenham, UK
I am speaking on behalf of the World Health Organization (WHO)
(regional office for Europe, Copenhagen). The Water Research
Centre, being a WHO collaborating centre, wishes to draw the
delegates' attention to WHO's interest in all aspects of
sewage reuse, particularly its effects on human health. I
refer to the WHO EURO Reports and Studies No. 42 (1981) which
reports on a WHO seminar on 'Health aspects of treated sewage
re-use' held in Algiers, 1–5 June 1980.
   This report discusses the principal forms of water reuse –
agriculture, aquifer recharge, industrial reuse and potable
reuse. The report considers that the last item may be of
greater importance than is generally realized if indirect
discharges to water supplies are included. The risk of
infection is considered the primary hazard of reuse.
   Chemical hazards are considered to be of less importance to
man but phytotoxicity is a significant consideration for
agricultural usage. The report suggests that treatment must
be adjusted to meet the needs of each specific situation and
that attempts to establish universal requirements for
treatment or water quality tend to be counterproductive. High
levels of treatment required to meet strict water quality
standards are very expensive and should only be employed where
they are needed.

MR D. D. YOUNG, Paper 1
Paper 16 in the session on health aspects gives a definitive
account of current knowledge on health matters. However, I
want to stress one point. Contrary to widely held belief, any
health issues which might be associated with organic compounds
are concerned far more with the naturally occurring substances
present in all surface – derived waters, as modified by
subsequent treatment, and that most studies looking for
specific toxic contaminants of industrial origin have given
reassuringly negative results. On the choice between high
reuse sources on the one hand and the taking of land for
reservoir construction or diminishing the base flow of rivers
on the other, no generalization should be made because the
choice will be influenced by the extent of reuse, land
requirement or stream diminution and by the cost in the
various alternatives. I am quite confident that, given modern
standards of treatment including processes such as activated
carbon, reuse alone shall never disqualify a potential source.

PROFESSOR B. Z. DIAMANT, Paper 2
In reply to the discussion, I wish to recommend strongly the
restricted reuse of sewage effluent.

MR A. G. STROM, Paper 3
In reply to Mr D. O. Lloyd, the reason for Australia's figure

of 4.6% reuse of effluent does not in any way lie with 'the division between the authority responsibilities within the country' but with the reasons outlined in the paper - chiefly that most of the population lives in well-watered cities on the seaboard. Reuse is proportional not to the size or nature of the organization, but to the hotness and dryness of the climate. Large comprehensive authorities such as the Sydney Water Board and the Brisbane City Council in fact have the lowest reuse. Also, it is not correct to say that Australia is a small user 'of water reclamation projects from sewage treatment plants'. The percentage of deliberate reuse in Australia is probably higher than in the USA or UK.

As to the use of computer modelling, streamflow data bases and models have been established but the great distances in Australia would preclude optimization of water reuse except on a regional scale. In the author's experience, political considerations usually confound any such well-meaning optimization.

In reply to Dr Swanwick, beef measles is certainly present in Australia; an inspection of cattle at slaughter in Victoria gives around 0.12% of animals infected. T. saginata prevalence in the population is very low, as hygiene standards are high and, importantly, beef is usually cooked well enough to destroy any viable cysts. However, this may be on the increase with immigration from the Middle East, in that a study (1978) into T. saginata infection in maternity hospital patients showed that Lebanese migrants accounted for some 89% of cases found. This is presumably due to the number of Lebanese dishes which include raw meat.

The 90-year-old Werribee Farm remains as one of the principal land treatment plants in the world and one of the very few left in Australia where raw sewage is still applied to land. No such practice would be permitted today in a new plant. Sewage is applied to 4300 ha of 8 ha paddocks divided into 0.16 ha bays, at intervals of 18-20 days over 6-7 months of the year. The bays are allowed 5-7 days to dry out before being grazed by cattle and sheep. Cattle must be bred on the farm and slaughtered under special inspection procedures, the incidence of animals found with living cysts being several times the state average. The long-term plan is eventually to upgrade the plant by first introducing primary sedimentation. Work has shown that other plants incorporating a measure of lagooning are quite free from T. saginata eggs.

It may be of interest that Dr Rickard et al. of the Melbourne University Veterinary Clinical Centre showed that calves could be effectively immunized against beef measles, but large-scale immunization appears unrealistic owing to the difficulties in obtaining adequate supplies of T. saginata eggs. Probably a cheap and effective vaccine will be found through genetic engineering, by developing bacteria to synthesize the T. saginata antigens.

With regard to the Frankston (Vic) trials of the growth of vegetables using chlorinated secondary effluent, reference 5 of the paper gives a most interesting summary of the work and should be compulsory reading for overseas practitioners in reuse. One finding of the study was 'under the conditions of this trial, there is no risk of viral infection from consuming vegetables irrigated with reclaimed water', the conditions including a chlorinated, high grade activated sludge effluent pumped from a storage dam. At present the Melbourne Board of Works is developing a scheme to pump this effluent some 15 km to a vegetable-growing area to augment the failing groundwater supply there. The health department has given its approval, under strict conditions, to the scheme, which includes the growth of vegetables. The conditions include 30 days' storage on site prior to use, and an extensive monitoring programme. What public reaction will be to all this remains to be seen.

4 Reuse of industrial effluent

R. C. SQUIRES, BTech, MIMechE, MIWES, MIWPC, Binnie & Partners

SYNOPSIS. The desire to control industrial pollution, the costs of developing the conventional water sources remaining after exploitation of the simpler, inherently-cheaper schemes located closest to the points of demand, and the crippling droughts apparently occurring more often and dwelling longer, have forced the countries in Southern Africa to undertake major research to identify the optimum strategy for implementing industrial water reuse without imposing severe financial burdens on these consumers. This paper shows how water reuse may be accomplished and provides data applicable to the meat, fishing, fruit and vegetable processing, tanning, mining and other industries.

BACKGROUND

1. Experience has shown that the imposition of water restrictions based on an overall percentage reduction in water used, causes severe difficulties to many industries owing to different standards of water and effluent management in the various undertakings. A survey initiated to examine the quality of water used and how it was used within the manufacturing processes, revealed that a wide variation occurred in water used from industry to industry and even between industries manufacturing the same product, often using similar process plant and techniques. A comparison of water usage between manufacturing concerns was made based on specific values obtained by dividing the water intake in kl by either the quantity of raw materials processed or by the weight of final product.

2. Table 1 shows the extent of variations in Specific Water Intake (SWI) and also the National Average Specific Water Intake (NASWI) derived by dividing the total water intake of the nation for a particular industry by the total national production of that commodity.

TABLE 1
Specific water intake (SWI) : Variation within industries

Industry		Units	Specific water intake	National average specific water intake(NASWI)
Abattoir:	Red meat	kl/cu[+]	1.2 - 5	2.5
	poultry	l/bird	8 - 50	20
Bread		kl/t	1 - 3	2.3
Breweries:	malt	kl/t	8 - 10	9.0
	sorghum	kl/t	2 - 4.3	3
Dairy:	milk	kl/t	2 - 4	-
	cheese	kl/t	10 - 30	14
	butter	kl/t	- 4	3.5
milk powder		kl/t	9.8 - 32	x
fruit & vegetable:				
	canning	kl/t	1.2 - 16	7.14
	freezing	kl/t	13 - 43	30
Soft drinks		kl/t	2 - 20	x
Vegetable oils & fats		kl/t	1.8 - 11	6.9
Tanneries		kl/t	8.2 - 10	x

+ cu = cattle unit - 1 cow,bull or ox, 5 pigs, 3 calves or
                    15 sheep
x = yet to be determined

3.   Prior to setting the target values for water intake,
initially at the NASWI, and thereafter at more stringent
values, detailed pilot plant trials were conducted within
the different groups of factories so that a Code of Practice
for water and effluent management within the industry could
be formulated showing how water reuse techniques could
reduce both water consumption and pollution.

INDUSTRIES
Fruit and vegetable industry (canning and freezing)
4.   Table 2 shows the leading parameters for processing
a wide range of fruits and vegetables. Inspection of the
table reveals that the quality of the final effluent
characterises them as relatively strong. Accordingly
rennovation of effluents for reuse must be selectively
applied. The approach to the problem is exemplified by
reporting the work conducted into peach canning.
5.   Fig 1 shows the distribution of effluent and its
quality expressed as percentages of the total along the
process line. It is immediately clear from Fig 1 that 80% of
the organic contamination resides in less than 50% of the
flow. Treatment and/or segregation of the effluents arising
from the pitting, peeling and washdown steps of the overall

TABLE 2

Commodity & Style		SWI Range kl/t	NASWI kl/t	Filtered COD kg/t	Filtered COD mg/l	SS kg/t	SS mg/l	PV kg/t
Apples	C	4.3–11.2	6.6	33.8	6800	2.0	430	3.2
Apples	J	1.8	1.8	10.7	3095	4.1	4148	–
Apricots	C	2.5–14.5	5.5	5.2	3646	5.5	390	11.2
Beans in tomato	C	20–70	20.0	–	2000	–	154	–
Beetroot	B	8–16	8.0	98	5940	4.4	270	18.0
Citrus	C + J	1.1–2.6	2.1	13.6	6500	–	–	–
Corn	C	6–11	9.8	6.3	2100	0.9	300	–
Green beans	C	7.4	7.4	8.2	1490	2.2	400	–
Guavas	C	4–10	6.4	3.2	700	0.9	195	–
Peaches	C	2.5–11.5	5.5	26	4544	5.0	880	5.7
Pears	C	4.5–12.9	12.7	26	2930	4.1	464	3.93
Pears	J	1.4–1.9	1.5	23.5	15000	7.7	4900	–
Peas	C	19–25	22.0	14.0	2000	0.8	120	–
Pineapples	C	2.1–4.5	2.94	18.0	12900	1.7	1190	–
Strawberries	C	6.8–27	17.0	–	–	–	–	–
Tomatoes	C	2–30	2.44	2.7	1100	1.2	493	–
Freezing								
Broccoli	F	8.1	8.1	49.0	5930	0.4	50	–
cauliflower	F	25	25.0	–	1950	–	274	–
Carrots	F	6–26	6.05	90.0	14980	28.5	4717	–
Corn	F	4.6	4.6	4.6	1000	0.5	100	–
Green beans	F	10–25	25.0	7.1	286	2.7	100	–
Peas	F	30	30.0	–	1950	–	274	–
Potatoes	F	25.7	25.7	74.0	2880	83.9	3266	–

Notes: Style – C = canning   J = juicing
              B = bottling   F = freezing

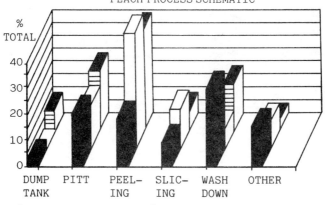

PEACH PROCESS SCHEMATIC

Fig. 1. Effluent quality ■ FLOW □ COD ▤ SS

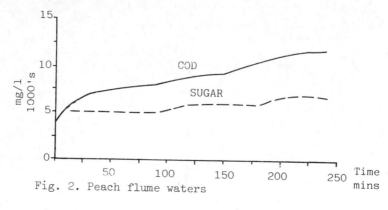

Fig. 2. Peach flume waters

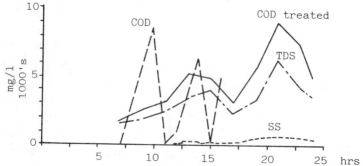

Fig. 3. Flume water reuse

Photograph 1. A plate and frame (centre) and tubular membrane test rigs operating on peach flume water after DAF treatment (para. 9)

process line ensures that the remaining effluents are relatively clear.

6.   Elimination of the pit removal flumes and their replacement by dry techniques, and the segregation of the peel sludges following caustic lye peeling allow the remaining effluents arising from the flume and the washdown to be examined as potentional sources for reuse.

7.   Fig 2 shows the effect on the COD and sugar concentration in mg/l of adding peeled half peaches to clean water. The rise in COD concentration is initially rapid due to leaching of fruit sugars from the newly-peeled fruit. As legislation governs the sugar content in canned peaches, the loss of sugars indicated by the graph represents the amount which must be made up by syrup addition prior to can closing; recovery of these sugars could represent a viable undertaking as about 4% of the natural fruit sugars are lost during processing; a plant processing 17 t/h of peaches could waste 300 kg sugar per hour which represents £30 per hour of operation or £9600 per month assuming 20 days operation at 16 hours per day.

8.   Flumes for transporting the partially-processed product are normally filled with potable water at the start of processing and dumped about every 3 to 4 hours when the organic load has risen sufficiently to discolour the water and affect the product quality. A dissolved-air flotation plant was connected to a typical peach flume. The plant consisting of a 2m dia reaction tank, a saturator recycle flow of treated effluent at 25% of the feed flow and saturator pressure 700 kPa, was fed with 8 kl/h of flume water. The plant maintained the SS load at a low value and substantially reduced the COD for about 10 hours. Fig 3 shows the SS and COD for both before and after the addition of the dissolved-air flotation plant. Prior to using the DAF plant, 3 to 4 fillings of each flume were necessary per processing day, saving about 400 kl/day.

9.   Once the flume waters have been treated sufficiently for reuse, the possibility of recovering the sugars in the flumes by ultra filtration and reverse osmosis exist. Fig 4 shows the typical results obtained from the RO trials conducted in the two test rigs shown in the photograph. The water recovered was patently of a quality for suitable reuse.

10.   Fig 3 also shows that the TDS of peach processing effluents rises significantly owing to the presence of NaOH from the lye peeling operation. The TDS must be reduced in this effluent as it is the main contributor to the final effluent TDS. Work is in hand using low rejection reverse osmosis membranes for separation of the caustic soda from the organics. The work already shows enrichment of the permeate alkalinity, meaning that some selective separation of the caustic soda is occurring.

11.   The plant washdown waters are treated by dissolved-air flotation removing the major portion of the SS. The

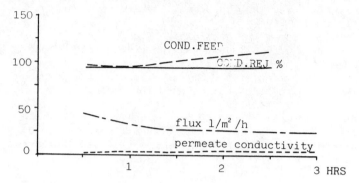

Fig. 4. Peach flume water treatment R.O.

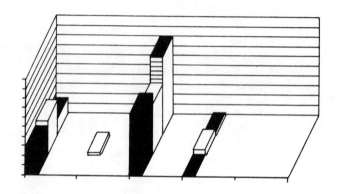

Fig. 5

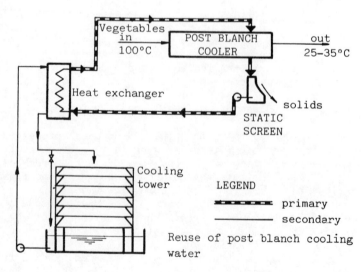

Reuse of post blanch cooling water

Photograph 2. Heat exchangers
for post-blanch cooling water
recycle loop (para. 13)

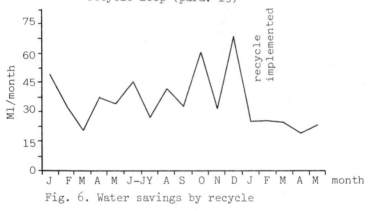

Fig. 6. Water savings by recycle

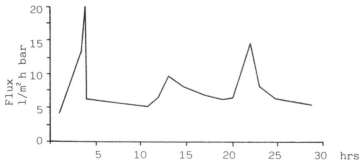

Fig. 7. Ultrafiltration of raw screened abattoir effluent

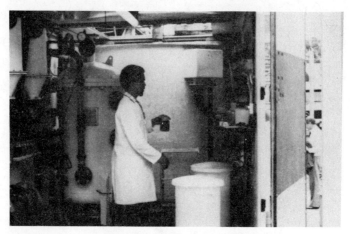

Photograph 3. Containerised pilot dissolved-air
flotation plant (para. 17)

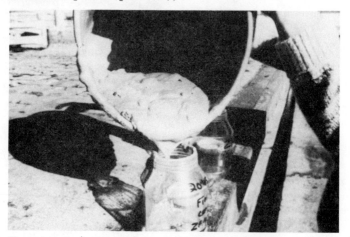

Photograph 4. The concentrate from the reverse osmosis
plant treating abattoir effluent (para. 20)

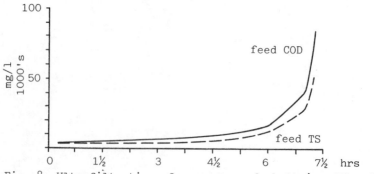

Fig. 8. Ultrafiltration of raw screened abattoir effluent

following management strategy evolves for implementing water
reuse in the fruit processing industry:-

- (i)   segregate pitting, peeling and first
  wash effluents;
- (ii)   treat (i) for solids removal (and later for
  caustic soda removal);
- (iii)   treat flume waters by DAF and recycle (later
  recovering useful sugars);
- (iv)   divide plant washdown into three stages as
  follows:
  - (a) gross solids removal by effluent
    segregated to exclude those in (i) above
  - (b) foaming antiseptic agent
  - (c) final wash with potable water.

Vegetable freezing effluents

12.   The processing stages used to prepare commodities for
freezing are similar to those for canning with the effluents
exhibiting similar characteristics. The majority of the SS
and COD being in the first washing, peeling and blanching
stages. However prior to freezing it is common practice to
reduce the temperature of the commodity from 100°C to about
25°C, either by water sprays or post blanch cooling flumes.

13.   The high water use of these post-blanch flumes can be
seen from the range of SWI reported at Table 1 viz carrot
freezing 6 to 26 kl/t and green beans 10-25 kl/t. Those
undertakings who recycle the post-blanch cooling waters
achieve a reduction in water intake of 10 kl/t. Fig 5 shows
shows the schematic diagram of the post-blanch cooling
system installed at one factory. It is based on a plate heat
exchanger removing the heat from spray coolers in individual
primary circuits and passing it to a common evaporative
cooling tower; several heat exchangers can be serviced by
one cooling tower. The photograph shows the minimum-space
needed for the heat exchangers. Counter current recyle was
also applied to the process lines.

14.   At the same plant the effluents have been segregated
so that those arising from the first three processing steps
are passed to a dissolved-air flotation plant for solids
removal and water meters have been installed to each process
line. The effects of the measures taken are immediately
apparent from Fig 6 which shows that since commissioning a
saving of over 17 Ml/month has been achieved which after
allowing for previous effluent disposal charges gives a
saving of about £9000 per month and avoids surcharges
previously levied at £20 000 per annum owing to the high
suspended solids concentration previously discharged.

Meat processing.

15.   Abattoir effluent is notoriously expensive to treat
employing current technology which favours pretreatment
followed by some method of biological degradation. Table 1
shows that the water intake varies from 1.2 to 5 kl/cattle
unit slaughtered, and values higher than this are recorded
at some older abattoirs. About 80% of the water intake is

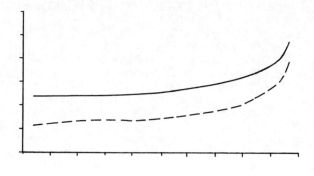

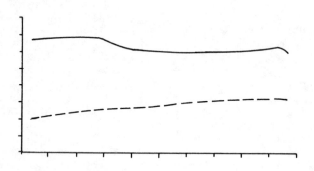

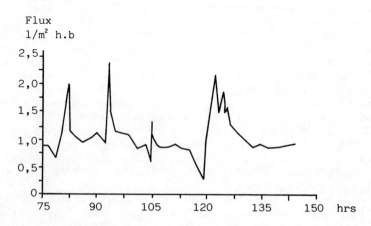

Fig. 9. Reverse osmosis - raw screened abattoir effluent

returned as industrial effluent with the remainder
comprising effluents from washing down of lairages and
domestic flows which passes to the sewer.

16.   The organic strength of the industrial effluent
varies widely depending on the methods of blood recovery and
paunch content handling. Typical figures are as follows:-

	C.COD(mg/l)	SS(mg/l)	TKN(mg/l)	TS(mg/l)
mixed abattoir	2500-5000+	1000-3000	300-500	2000-3000*

+ blood discharges can provide values up to 20 000
* blood discharges can provide values up to 10 000
C = centrifuged

17.   A containerised dissolved-air flotation pilot plant
operating at a surface loading of $2.0m^3/m^2/hr$, treating
balanced, screened effluent dosed with acid to pH2 and 120
mg/l sodium hexameta-phosphate, produced a good quality
effluent with SS concentration of about 500 mg/l and a COD
concentration of about 1000 mg COD/l. However the
variability of the influent feed quality, such as occurs
when the mucus from paunch opening escapes into the
drains, causes instability which results in periods of floc
carry over with consequent deterioration of the effluent
quality. The sludges recovered by this protein precipitation
process amount to about 750 kg 'dry sludge' (10% moisture
content) per 1000 kg COD removed in the process. The sludges
arising from treating the effluents from a mixed abattoir by
such processes have about 17% fat content together with
protein and a small proportion of ash.

18.   The sludges recovered from the plant contain about 5%
solids. They can be dewatered by decanters after pH
correction and heat coagulation before being passed to a
rendering plant with other inedible material to form carcass
meal. The revenue from this source can often provide 17%
return on capital invested.

19.   The treated effluents from DAF processes however,
still require further upgrading to potable quality to adhere
to regulations governing water usage in an abattoir; severe
standards being imposed on water quality in this industry by
the veterinary licensing authorities.

20.   As a result of this and of the possibility of
inferior effluent quality arising from instabilities in
chemical treatment, research has been conduted over the last
12 months at an abattoir to determine the benefits of
employing membrane processes to achieve the desired quality
of water reuse while producing a drier concentrate for
recycling as animal feed.

21.   The first series of trials involved ultra filtration
using tubular non-cellulosic membranes operating at
pressures between 4 and 10 bar on the industrial (faecal
free) effluents after screening through a wedgewire screen
of 1mm apperture. The trials have been operated as batch
recycle trials, with each concentration cycle terminated

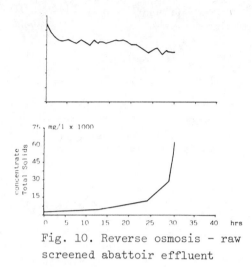

Fig. 10. Reverse osmosis - raw
screened abattoir effluent

Photograph 5. Plate and frame
membrane unit operating on
stickwater at 80°C (para.25)

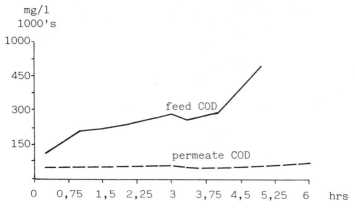

Fig. 11. Stickwater concentration UF

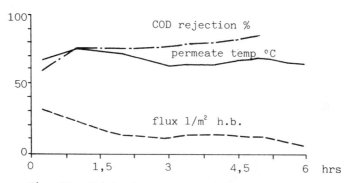

Fig. 12. Stickwater concentration UF

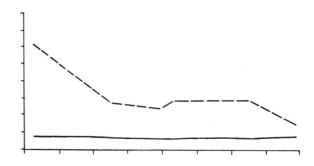

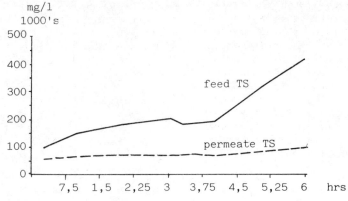

Fig. 13. Stickwater concentration

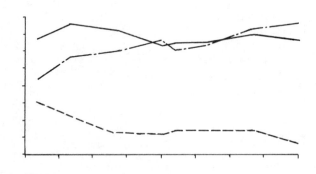

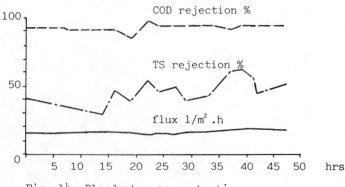

Fig. 14. Bloodwater concentration

when the Total Solids (TS) reached about 8-10% TS. The
cleaning regime has now been established and typical results
of flux expressed as $1/m^2$ h.bar are given in Fig 7; the peaks
represent the restoration of flux after cleaning. Fluxes in
excess of 5 $1/m^2$ .h.bar or 40 $1/m^2$ .h are readily achieved.
Fig 8 shows that the total solids concentration in the
concentrate reached 8%, with permeate qualities of less than
1000 mg COD/l and 750 mg TS/l. The plant operated without
difficulty with varying feed (abattoir effluent) quality.
22.   Trials were then conducted with non-cellulosic tubular
reverse osmosis membranes. Fig 9 shows that after 150 hrs of
batch recycle operation fluxes approaching 1.0 $1/m^2$. h.bar
(35 $1/m^2$ .h) are being achieved; at the time of writing over
350 hrs of similar operation has been accumulated.
23.   Fig 10 shows the total solids concentration achieved
during a typical 30 hr run, during which an average flux
of about 35 $1/m^2$ .h was achieved, the total solids of the
concentrate reached 6.5% TS and the permeate COD remained
below 50 mg COD/l for most of the trial. The permeate total
solids concentration remained at less than 20 mg TS/l for
most of the trial. Bacterial analysis of the permeate shows
them to be acceptable for reuse. The solids are reused as
animal feed.
24.   Work is continuing to accumulate a further 12 months
operation at a continuous concentrate concentration of 6%
total solids.

Fishing industry
25.   Fishmeal manufacture and fish canning, produce strong
organic effluents many of which are presently passed to the
ocean, representing both loss of valuable products as well
as pollution. Rennovation of these waters to recover the
useful protein and fats and reuse of fish tranportation
flume waters has been addressed by pilot membrane trials
using tubular and plate and frame configurations employing
non-cellulosic ultra-filtration membranes. Reuse of certain
liquors derived from processing of fishmeal is common
practice, the most notable of which is the stickwater
derived from the pressing of the fish after cooking. These
liquors are normally evaporated to 35% total solids, the
condensates are available for reuse and the concentrates
pass to the fishmeal driers to become product.
26.   Figs 11 to 13 show typical operating results for the
plate and frame units operating on stickwater. Permeate
fluxes of 15 to 20 $1/m^2$ .h have been repeatedly achieved and
the total solids concentration of the fluid has been raised
to over 40% TS. Careful control of the operating pressure at
about 4-6 bar and temperature is needed for repetitive
results. Further trials are in progress with membranes of
different cut-off molecular weights.
27.   Trials using tubular membranes operating on
bloodwater generated during unloading of the fish from the
fishing vessels have also been undertaken. Figs 14 and 15
show that at ambient temperature 8% Total Solids

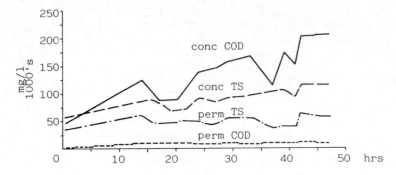

Fig. 15. Bloodwater concentration

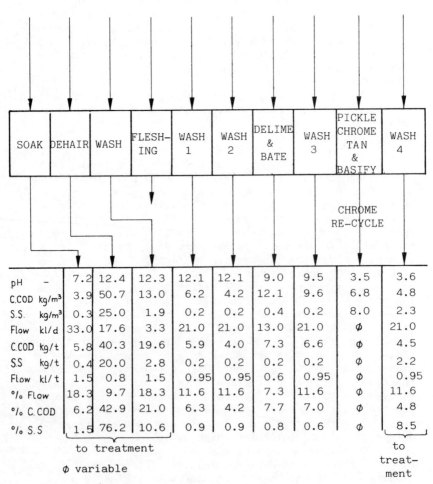

		SOAK	DEHAIR	WASH	FLESH-ING	WASH 1	WASH 2	DELIME & BATE	WASH 3	PICKLE CHROME TAN & BASIFY	WASH 4
pH	–	7.2	12.4	12.3	12.1	12.1	9.0	9.5	3.5	3.6
C.COD	kg/m³	3.9	50.7	13.0	6.2	4.2	12.1	9.6	6.8	4.8
S.S.	kg/m³	0.3	25.0	1.9	0.2	0.2	0.4	0.2	8.0	2.3
Flow	kl/d	33.0	17.6	3.3	21.0	21.0	13.0	21.0	⌀	21.0
C.COD	kg/t	5.8	40.3	19.6	5.9	4.0	7.3	6.6	⌀	4.5
S.S	kg/t	0.4	20.0	2.8	0.2	0.2	0.2	0.2	⌀	2.2
Flow	kl/t	1.5	0.8	1.5	0.95	0.95	0.6	0.95	⌀	0.95
% Flow		18.3	9.7	18.3	11.6	11.6	7.3	11.6	⌀	11.6
% C.COD		6.2	42.9	21.0	6.3	4.2	7.7	7.0	⌀	4.8
% S.S		1.5	76.2	10.6	0.9	0.9	0.8	0.6	⌀	8.5

to treatment   to treat-ment

⌀ variable

Fig. 16. Tannery effluents

concentration can be achieved, enabling the concentrates to be reused in the fishmeal manufacturing process. The permeates are generated at a flux of about 15 l/m^2.h and have a COD concentration about 10 000 mg COD/l when the concentrate COD approaches 200 000 mg COD/l. Reuse of these permeates for fish transportation is being examined.

28. Inspection of the operating results demonstrates the importance of a simple membrane cleaning regime. Flux restoration at the completion of each run is being successfully accomplished.

Tanneries

29. Tannery effluent is particularly difficult to dispose of, although biological treatment can be successfuly accomplished. Fig 16 shows the character of the nine major effluents which make up the discharge from a wet-blue tannery; i.e. a tannery which does not finish leather but produces a wet chrome tanned product. It can be seen that less than 58% of the effluent contains nearly 75% of the centrifuged COD load of the entire tannery and 97% of the suspended solids.

30. The lime sulphide liquors from the first three process steps contain large quantities of lime-based solids which settle readily. These liquors are therefore settled in a vertical flow settlement tank. The only other liquor warranting treatment is the chrome wash liquor (wash 4), assuming chrome recycle is practised. Again these liquors settle readily.

31. After the first stage settlement the liquors are re-mixed, passed through a flocculator and introduced into a third settlement tank where auto precipitation occurs. Aeration of the clarified liquor removes the residual sulphides before the final treatment stage is accomplished by addition of FeCl$_3$ followed by either settlement or dissolved-air flotation.

32. The first settlement of the lime sulphide liquor reduces the SS from 19 000 mg SS/l to about 4800 mg/l with a consequent reduction in the centrifuged COD from 49 000 mg/l to 33 000 mg COD/l. The effluent from the mixed liquor stage has a centrifuged COD concentration of 11 500 mg/l and an SS concentration of 400 mg/l. The final treatment step by DAF reduces the C.COD to a concentration of 2200 mg/l with a corresponding SS of 150 mg/l. If settlement is used in place of DAF, the final effluent quality will have a COD concentration of 2700 mg/l and an SS of 2015 mg/l.

33. The sludges recovered from the various treatment stages were tested for their amenability to dewater naturally on open drying beds. Fig 17 shows that the lime sulphide sludges alone required 10 days to achieve 19% solids, whereas if they were mixed with the sludge from the final treatment stage, 40% solids could be achieved in 10 days without smell from the drying sludges.

34. Recovery of waters for reuse is governed by the TDS, fat and protein content of the water; accordingly the pre-

Photograph 6. Tannery pilot treatment plant for selective
settlement of wet blue effluents (para. 32)

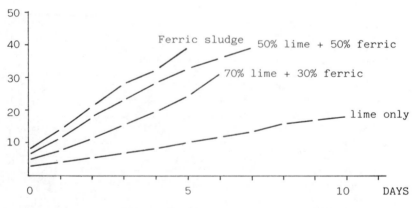

Fig. 17. Tannery sludge drying

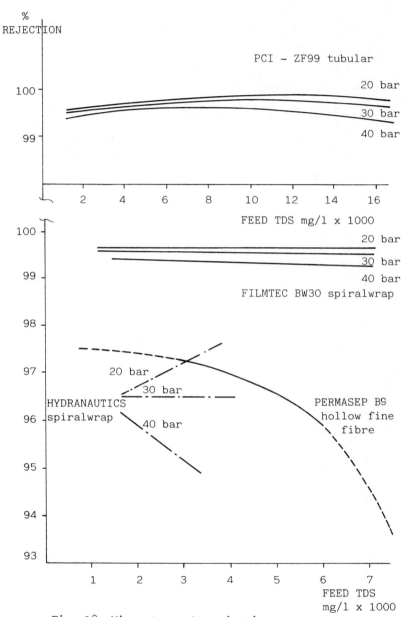

Fig. 18. Minewater salt rejection

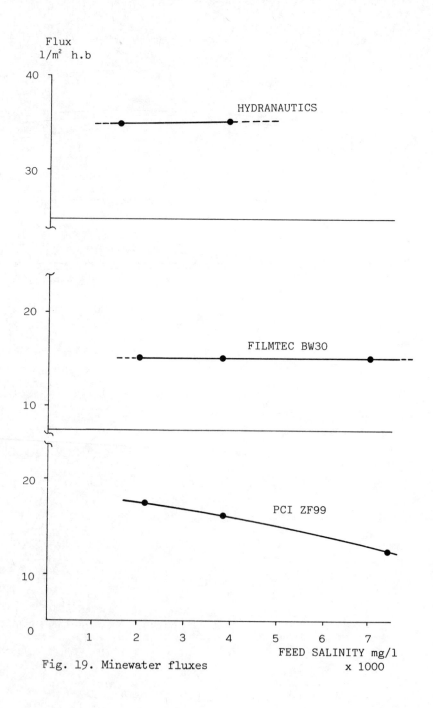

Fig. 19. Minewater fluxes

treatment outlined herein permits the desalting of the effluent prior to reuse.

Mining and other effluents

35. Mining operations at a deep bituminous coal mine required the pumping of 4000 m^3/d of drainage water to the surface for disposal. The salinity of this effluent is up to 2000 mg/l of which 500-800 mg/l is present as sodium, mainly in association with the bicarbonate ion. Owing to a restriction in the sodium permitted to be discharged in the area, treatment by reverse osmosis is necessary with the permeate being reused as low pressure boiler feed water.

36. The characteristics of the mine water were determined to be:-

Parameter		Average value	Parameter	mg/l
pH	–	8.5	sodium bicarbonate	816
Conductivity	μS/cm	2375	sodium sulphate	141
TDS	mg/l	1720	sodium chloride	190
SS	mg/l	8	calcium bicarbonate	15
Turbidity	NTU	1.5	magnesium bicarbonate	13
Temp	°C	35	sodium fluoride	10
			potassium chloride	7
			potassium nitrate	1
			silica	10
			TDS	1203

37. Owing to the high sodium bicarbonate concentration, it was elected to attempt recovery of bicarbonate from the concentrate, so that soda ash could be manufactured for use on the site in other processes. A pilot plant was erected to investigate the feasibility of high water recovery without acid pretreatment which would destroy the bicarbonate and hence the by-product. As the calcium/magnesium hardness was low, ion exchange units were used to eliminate these elements. A small scale single membrane rig was used to explore the performance of various membrane types.

38. Fig 18 shows the varying rejections of TDS by the membranes of 4 different manufacturers and Fig 19 shows the fluxes obtained when operating on the effluent. The Filmtec BW was selected for the large pilot unit.

39. The pilot plant shown in the photograph was designed to demonstrate a water recovery of 90% with a once-through passage of mine drainage water in a number of membranes arranged in series/parallel taper 4:2:1. Stage 1 used 4 parallel lines of 4 membrane elements, stage 2 two lines each of 4 elements and the final stage used a single line of 4 elements. The plant was fed with pretreated water at the rate of 26 l/m (37 m^3/d).

40. Pretreatment was accomplished using prechlorination for algae control, coagulation by dosing 2 mg/l of polyelctroyte, dissolved air flotation, filtration and ion exchange. The Silt Density Index of this water was reduced from in excess of 100 to between SDI 4 and 5. Scale control

Photograph 7. Reverse osmosis
plant treating mine water
(para. 39)

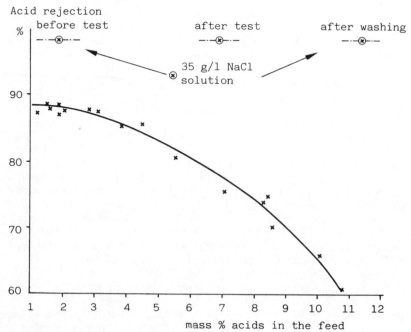

Fig. 20. Aliphatic acid concentration

was performed using Flocon 100 at a dose of 5 mg/l. The permeate produced at 90% water recovery had the following analysis making it suitable for reuse as boiler feed water:

	mg/l	
NaCO	46	(alkalinity 28 mg/l as $CaCO_3$)
Na_2SO4	8	
NaCl	26	
TDS	80 mg/l	

The total hardness was less than 0.25 mg/l (as $CaCO_3$) and the soluble silica about 1.1 mg/l.

The concentrate produced had the following analysis:

	kg/1000 kg water
NaHCO3	16.58
Na_2SO4	2.97
NaCl	3.25
sol Silica	0.10

The permeate fluxes obtained at 30 bar pressure were 15 l/m^2. h.bar.

Acid concentration by reverse osmosis

41.   An effluent stream from a petro-chemical works contained about 1.3% of mixed aliphatic acids including acetic, proprionic and butyric acids; the major proportions being acetic. Other compounds present included phenol, methanol and traces of higher alcohols, aldehydes and ketones. These compounds were being passed in the effluent to a biological treatment plant where certain of them were thought to be causing sludge bulking. Accordingly removal of the acids was needed.

42.   Trials were performed using reverse osmosis to concentrate selectively the mixed acids up to about 11% concentration. Fig 20 shows the results of the trials depicting acid rejection as a function of the % by mass of the acids in the feed. As the test unit operated on a batch concentration with concentrate recycle, the feed·was continually concentrated until the flux reached its final value. Nearly 11% by mass of acids was achieved during the test runs.

43.   The permeate rate fell from 112 ml/min to 10 ml/min during the trial while the permeate conductivity rose from 200 μS/cm to 1500 μS/cm as the concentration in the feed increased from 1 to 11%.

44.   Reverse osmosis for selective concentration of the acid stream is demonstrated and the permeates can be recovered for process water duty.

Seeded slurry reverse osmosis

45.   Many underground mine waters are saturated with calcium sulphate and these have successfully been desalted without pretreatment in tubular reverse osmosis membranes employing the introduction of a calcium sulphate seed similar in principle to that used in seeded evaporators.

Over 5000 hours running has proved the process to have considerable merit.

Wool washing effluents

46.   Work conducted using a new-generation dynamic-membrane ultrafiltration process reveals that without pretreatment these membranes which can be regenerated in-situ can treat the raw effluent from the wool scouring and desuinting operations at temperatures approaching 90-95°C. The permeates recovered are satisfactory for reuse in the wool scouring line. Full closed cycle operation now appears feasible.

Crossflow microfiltration

47.   New developments in this relatively low-pressure process, employing replaceable membrane layers, reveal its potential for concentrating alum sludges, in rennovation of certain paper mill effluents for recycle and as a pre-treatment process prior to reverse osmosis, especially for recovery of potable quality water from secondary sewage effluents.

Recovery of secondary sewage effluents by reverse osmosis

48.   Secondary sewage effluent from an activated sludge sewage treatment plant is filtered in sand filters, chlorinated below breakpoint and stored before being pumped through 50 μmm cartridge filters to tubular cellulose acetate membranes. Over 6300 hours of operation have been successfully accomplished.

49.   The pilot plant is producing an average product recovery of 75%. The salt rejection using a standard 2000 mg/l solution of sodium chloride is checked periodically and after 3300 hours it was 86.7%. The plant overall conductivity rejection remains at 89-91% which has not decreased over a 5000 hour period.

CONCLUSIONS

50.   Advances in membrane technology are now occurring very rapidly; however while their capabilities for desalting brackish to seawater are well documented, little data are recorded on their performance with effluents. If they are selectively applied to individual effluent at their source within the manufacturing process, they could perhaps be the key to treating industrial effluents for re-use while recovering valuable materials from the concentrates to offset the capital and running costs while both conserving water and reducing pollution.

5 Reuse of sewage effluent in industry

M. R. G. TAYLOR, PhD, DIC, CChem, FRSC, MIWPC and J. M. DENNER, MA, MSc, DIC, MICE, MIPHE, Balfour Consultants in Environmental Sciences Ltd, UK

SUMMARY

1. Treated sewage effluent is re-used by industry but its use is generally restricted to low grade applications where no constraints are imposed by hygiene or special water quality requirements. The demand for sewage effluent re-use in various sectors of industry and the technical and economic aspects of re-use, quality requirements for the different industrial applications and operational problems are discussed. Specific case studies around the world are reviewed.

INTRODUCTION

2. Considerable volumes of water are used by industry for process purposes and for cooling. In 1970 it was estimated that $1.5m^3$ of water were required in the production of 1 tonne of coal, $90m^3$ for a tonne of paper and $200m^3$ for a tonne of steel(1). The demand that this consumption makes on available water resources, coupled with the initial cost of water and the difficulties and cost of disposal of the associated waste water, has led to greater attention to the need for more efficient use of water in industry. In recent years much more attention has been given to conservation, recycling and re-use, and as a result the amount of water used per unit of production in industry has decreased substantially. In addition, alternative sources of lower grade water have been found to be acceptable in many industrial applications.

3. One of the most important alternative sources of water is municipal sewage effluent. In terms of quantity it can be the most stable and dependable source of supply in areas where water resources are limited. However, in considering the re-use of sewage effluent in industry, general aspects of water management, such as the need to maintain river flows, recharge aquifers, provide potable water and meet agricultural demands have first to be taken into account to ensure the optimum use of the resource.

Reuse of sewage effluent. Thomas Telford Ltd, London, 1984

4. Water of any desired quality can be produced from
municipal sewage effluent by appropriate treatment, and
normally some degree of pretreatment is required before
sewage effluent can be used in industry. Inadequate water
quality can cause problems in respect of three major
categories of concern to the industrialist (2):-

a) Product degradation

    (i)   decay, resulting from biological activity
    (ii) staining or tainting
    (iii) corrosion
    (iv) chemical attack and/or contamination

b) Deterioration of plant and machinery

    (i)   corrosion
    (ii) erosion
    (iii) scale deposition

c) Reduction of efficiency or plant capacity

    (i)   tuberculation
    (ii) sludge formation
    (iii) scale deposition
    (iv) foaming
    (v)   biological growths

5. Treatment can be expensive, and the cost of producing
water of high quality may be prohibitive. By assessing the
actual water quality requirements at each stage of a process
it should be possible for the supply to be tailored to
satisfy the needs at each phase and thus avoid the expensive
treatment of all the water to meet the standard required for
the most critical use. This approach will contribute to the
optimum use of sewage effluent as a source of water for
industry.

6. Re-use of sewage effluent by industry has been reported
in many countries including the UK, USA, South Africa, Saudi
Arabia, Japan, Israel, Singapore, Mexico and the USSR. Where
effluent quality does not meet the requirements of industrial
users, further treatment is carried out, either at industrial
water works as in the case of the Kohtoh works in Japan (3)
and Jurong works in Singapore (4) or on-site by the user as
in Odessa, Texas (5).

7. The demand for re-use of sewage effluent depends on the
nature and size of the industry. Unless water resources are
very limited and cost considerations are therefore secondary,
the main use will be for cooling and low-grade process consumption, where water quality requirements are less critical.

Re-use in such industrial sectors as food, drink and pharmaceuticals is usually avoided for reasons of hygiene. It is of interest to note that the use of reclaimed water from the Jurong works by the food industry is prohibited (6).

8.   Table 1 details the degree of effluent re-use for various purposes in the USA, indicating that the major application for effluent re-use is for cooling water which is the largest specific use of water throughout industry.   A similar pattern of use may be seen in Table 2 for the Kohtoh Industrial Water Works in Japan which serves a total of 373 industrial concerns.

Table 1 - Industrial Re-use of Sewage Effluent in the USA.(3)

Type of use	Number of Plants	Re-use volume m^3/d	Percent of totals
Boiler feed	3	3 785	17
Cooling	12	582 890	66
Process	12	3 785	17
TOTAL	27	590 460	100

Table 2 - Usage of Water Supplied by the Kohtoh Industrial Water Works (1975)(3)

Use	Volume (m^3/d)	%
Cooling	52 162	50.7
Washing and Rinsing	31 338	30.4
Process	12 440	12.1
Miscellaneous	5 756	5.5
Air-conditioning	619	0.6
Raw Material	600	0.6
Boiler Feed	37	0.1
TOTAL	102 952	100

9.   Three main areas of re-use of sewage effluent in industry can be defined in order of decreasing volume as follows:

Cooling water	: medium quality
Low grade process use	: low quality
Medium-grade process use	: high quality

COOLING AND BOILER FEED WATER

10.   The major users of cooling water and therefore of sewage effluent are thermal power stations and the heavy industry sectors such as iron and steel and other basic metals production. In these industries there is also a major demand for steam, and in some instances sewage effluent has been used as boiler feed water after a considerable degree of treatment. Table 3 shows the quality criteria appropriate for cooling water and boiler feed water.

11.   The problems associated with the use of sewage effluent in cooling systems are biological slime growth, scaling, corrosion and foaming. Cooling water, whether used on a once-through basis or in recirculation systems, must not contain contaminants to the extent that they could cause blockages, for example by deposition in the cooling system. Build-up of soluble contaminants in recirculating cooling systems can be controlled by regular wastage, known as blowdown, and make up with fresh water.

12.   Gelatinous biological slimes occur in cooling systems as a result of excessive growth of bacteria, fungi and algae, stimulated by high nutrient concentrations or by the intense aeration and high temperature conditions obtaining in cooling towers. These slime deposits can seriously reduce the rate of heat transfer, and some bacterial deposits can be corrosive.   Excessive growth can be controlled by chlorination or by dosing the circulating water with slimicides.

13.   The most important problems in cooling systems are scaling and corrosion, with calcium, phosphate and alkalinity being the main scale-forming agents in sewage effluent. Under conditions of high pH insoluble phosphates are deposited on heat-exchange surfaces, resulting in loss of operational efficiency. However, where alkalinity and total hardness levels are low, with consequent low pH, potentially corrosive conditions can occur.

14.   The stability of sewage effluent depends on the hardness and alkalinity of the carriage water and on the nitrogen and phosphorus concentrations. During the aeration that occurs in cooling towers there is oxidation of nitrogen

species to nitric acid, which results in a reduction in alkalinity: oxidation of 1 mg/l ammonia-N to nitric acid reduces alkalinity by 7.14 mg/l as $CaCO_3$. Careful control of pH is necessary in the case of non-nitrified or partially nitrified effluents, particularly where the original water supply is of low alkalinity.

Table 3 - Water Quality Criteria for Cooling
        and Boiler Feed Use (7)

Parameter	Cooling Water		Boiler Feed Water
	Once-through	Recirculation Make-up	
Alkalinity	500	350	350
Aluminium	b	0.1	5
Bicarbonate	600	24	170
Calcium	200	50	b
Chloride	600	500	b
Copper	b	b	0.5
COD	75	75	5
Hardness	850	650	350
Hydrogen Sulphide	–	b	b
Iron	0.5f	0.5	1
Magnesium	b	b	b
Manganese	0.5	0.5	0.3
Nitrogen-Ammonia (asN)	b	b	0.1
Oil	no floating	–	–
Organics, CTE	g	b	1
MBAS	b	b	1
Oxygen, dissolved	present	b	2.5
pH, units	5.0 – 8.3	b	7.0-10.0
Silica	50	50	30
Suspended Solids	5 000	100	10
Total Dissolved Solids	1 000	h	700
Sulphate	680	200	b
Temperature °C	b	b	b
Zinc	b	b	b

Notes:

b   Accepted as received
f   0.5 mg/l iron and manganese
g   No floating oil
h   Effluent TDS values are typically in the range 500 to 800 mg/l.

15. Nitrification in cooling towers has, however, been used to advantage in the case of the Croydon power station, where a sewage effluent derived from a hard water of high alkalinity and containing significant concentrations of residual ammonia was used as the major source of water for cooling purposes (8). Severe problems of calcium phosphate scaling had been experienced, necessitating extensive condenser tube cleaning and acid washing, and these had been overcome at relatively high cost by chlorination to reduce pH. It was calculated, however, that nitrification of the residual ammonia would reduce alkalinity by 80 per cent of the amount required to prevent phosphate precipitation but was being prevented by the addition of chlorine. The mode of operation was changed to allow nitrification to occur and the required alkalinity and pH levels in the recirculating water were achieved by the addition of the requisite further amounts of ammonium sulphate. This method of operation obviated the need for the expensive manual cleaning of condensers and acid washing without causing problems of metallic corrosion.

16. In the USA sewage effluent has been used for cooling water at many power stations. In West Texas, an area of limited water resources, sewage effluent has been found to be the most economical source of water for cooling and boiler feed water in some thermal power stations and petrochemical plants (9). In addition, there is a further benefit in terms of pollution control in that such reuse provides a means of protection of receiving waters by removal and concentration of pollutants and relocation of disposal points.

17. In 1979 the Southwestern Public Service Company used an average of 19 tcmd of sewage effluent from the City of Lubbock, Texas for cooling water and boiler feed water in its power station (5). The effluent used for cooling water was given lime treatment to remove phosphorous and alkalinity and to provide some coagulation to reduce the organic content. For boiler feed water make-up the effluent was treated by anthracite - coal filtration, reverse osmosis and ion exchange using cation, anion and mixed-bed resins. Few problems have been experienced in the operation, and the blowdown from the cooling towers and the boilers is diluted with wastewater to reduce the concentration of total dissolved solids to about 2 500 mg/l and then reused for irrigation on local farms.

18. In the Odessa area of Texas a large refinery and petrochemical plant owned by the El Paso Products Company uses the entire sewage flow from the city of Odessa (9). The water is used for fire protection, cooling water make up and, after lime treatment, coal filtration, ion exchange and degasification it is also used for low pressure boiler

feed water. Water for high pressure boilers is deionised by strong acid and strong base ion exchange. The wastewater from the petrochemical plant is then injected into oil-bearing formations to increase the extraction of oil. This obviates discharges to watercourses, and in this way the quality of the water available to downstream users is protected.

19. In Arizona a nuclear power station uses secondary sewage effluent transported by pipeline from Phoenix for cooling water for three turbine generators (10). The flow of 340 tcmd (90 mil gal(US)/d) is treated by biological nitrification followed by two stage lime-soda softening and dual media filtration. The whole process has a computerized control system with flow equalization provided upstream by in-pipe storage and downstream by a treated-water reservoir.

LOW-GRADE PROCESS USE

20. In numerous industrial operations water is not involved in any chemical processes, but serves only as a carrier for the transport of raw materials, products or by-products.

21. The primary metals industry is a major user of sewage effluent. Treatment to reduce the suspended solids content may be necessary for use in cold rolling and reduction mill waters, but secondary effluent can be used directly in coke and slag quenching, gas cleaning and hot rolling processes. The largest industrial user of sewage effluent in the USA for many years has been the Sparrows Point plant of the Bethlehem Steel Company at Baltimore, where about 500 tcmd of sewage which has received limited biological treatment is used primarily for direct cooling of steel and slag (11).

22. In the UK, Commonwealth Smelting Ltd (CSL) at Avonmouth uses the effluent from the activated sludge plant at Bristol sewage-treatment works for gas cleaning, cooling and quenching molten slag in an integrated zinc and lead smelter plant (12). The decision to use effluent from the sewage treatment works, which is located less than 1 km from the plant, was prompted by the lack of other suitable water resources in the area to meet the demand of 9 - 14 tcmd for these uses. The effluent is chlorinated at the sewage treatment works and pumped into a holding reservoir on the CSL site, but no further treatment is undertaken. Sewage effluent is also used in the wet-process phosphoric acid plant for fume scrubbing and then re-used for conveying gypsum by-products as slurry through a 1.5km plastic pipe into the Severn Estuary. It is understood that no operational problems with the use of sewage effluent have been experienced.

23. In Saudi Arabia restricted water resource availability dictated that a petroleum refinery had to use sewage effluent as a raw water supply (13). Three levels of water quality were required: low grade for firefighting water, medium grade for desalters and cooling water; and high quality for high-pressure boiler feed. Extensive treatment was required for the production of high quality water, and a process scheme involving reverse osmosis, ion exchange and carbon adsorption was selected.

24. A sequential use of effluent at a coal-fired power station has been reported (14). The effluent was used for cooling water and the cooling water blowdown, then re-used for coal-pile spraying and make up water for flue gas desulphurization scrubbers.

25. Sewage has been used for transport of mine tailings and coal slurry transport using sewage effluent has been under consideration in the USA (15).

PROCESS WATER

26. For water use within the actual production process the requirements for water quality are usually more stringent than for cooling or cleaning as discussed previously. As a result secondary sewage effluent may not be acceptable, and it may be necessary to provide further treatment to remove suspended solids and possibly other soluble contaminants such as colour.

27. At the Jurong industrial water works, secondary effluent from the Ulu Pandan sewage-treatment works is further treated by prechlorination, alum coagulation and clarification, dual media (sand and anthracite) rapid gravity filtration, cascade aeration and post chlorination. The final effluent contains high concentrations of ammonia and dissolved solids, the latter originating principally from the water supply. Despite the need for this additional treatment, by 1976 a volume of some 23 tcmd was being used by a total of 43 industrial concerns out of some 675 industries in the Jurong Area (4).

28. The pulp and paper industries in Jurong use some 71 per cent of the output of the industrial water treatment plant. Large quantities of water are required in pulping and in transporting fibres through the paper-making processes. Operational problems can be experienced in the use of sewage effluent in paper-making. Organic matter and nutrients such as phosphate can cause clogging of equipment and slime growth; colour is adsorbed by the cellulose; hardness reacts with resins and interferes with bleaching and sizing of bleached paper; and heavy metals catalyse the decomposition

84

of calcium hypochlorite bleach liquor or are adsorbed onto
the cellulose, affecting the brightness. In spite of these
potential problems there is significant use of sewage
effluent by the paper industry in South Africa and elsewhere.

29.   In the wool textile industry high quality water is
usually demanded, but in some processing applications sewage
effluent has been used successfully. At Pudsey, Yorkshire,
at a mill producing all-wool blazer cloths a wet processing
sequence involving scouring with soap and soda, piece-dyeing
and milling was operated using sewage effluent as the sole
water source (16). Other sources of water supply to the mill
were a private borehole, a local stream and the mains
supply. The private borehole yielded water of excellent
quality but with diminishing output, and the water quality in
the stream was unsuitable at times of high flow because of
excessive suspended solids concentration. Thus in the event
of failure of these two sources it became necessary to use
mains water, which significantly increased process costs. In
view of these limitations it was decided to carry out
laboratory and plant-scale trials using secondary sewage
effluent.

30.   The secondary effluent was further treated by rapid
gravity sand filtration and then chlorinated at doses of up
to about 10 mg/l to achieve a residual after 1 hour of
1 mg/l. Initially it was thought prudent to remove residual
chlorine by addition of sodium thiosulphate before using the
effluent with particular dyestuffs. However, subsequent
experience showed that in general residual chlorine
concentrations were not at levels that would cause
interference with the dyestuffs normally used at the mill,
and dechlorination was therefore suspended. Following the
successful operation of this process sequence the supply of
sewage effluent was extended to a second dyehouse in which
cotton staining was being carried out.

31.   There is also extensive use of sewage effluent in the
textile industry in Jurong, accounting for some 20 per cent
of the total volume of effluent re-used.

ECONOMICS

32.   Substantial volumes of sewage effluent will only be
used by industry if there is a significant cost advantage in
so doing. In addition industries with low water consumption
are often unwilling to risk using sewage effluent in view of
the small cost savings. Consideration of the charges made
for effluent and the treatment costs in isolation can be
misleading, because equally important are the costs and
availability of alternative supplies. Thus in areas of very
limited sources of water of suitable quality it is still

economically attractive to use sewage effluent for high-grade uses in spite of the significant treatment costs.

33. Where sewage effluent is supplied to a single user the supply is governed by individual agreements, but it is normal practice for the user to be charged the additional treatment costs incurred at the sewage-treatment works for treatment provided to the effluent that would not otherwise be necessary (such as chlorination). In addition there would be a charge for any pumping required to transfer the effluent to the user's premises. In effluent reclamation plants serving a multiplicity of users charges are based on production costs and on the volume of water supplied and are typically in the range of 10-25 per cent of the cost of mains water.

34. In the case of the Jurong works the Singapore Government gives cost incentives to industry to use the reclaimed effluent which for large users (more than 50 000 m^3 per month) is 12 per cent of the mains water cost (4). In Moscow, tertiary sewage effluent is further treated by sand filtration and disinfection and is distributed to a number of industrial users, including car production and chipboard manufacturing plants, at a cost of 20 per cent of that of mains water (17).

35. A novel agreement, thought to be unique in the area, was made between the City of Odessa, Texas and the El Paso Products Co. The company operates its large oil refinery and petrochemicals complex in the Odessa area primarily because of the availability of large quantities of oil and natural gas feedstocks, but also because water resources are limited and the only dependable supply of water available is the sewage effluent from the city. The company therefore agreed to build, operate and maintain a secondary sewage treatment plant at no cost to the city in exchange for the use of all the sewage effluent.

36. An energy conservation scheme has been developed in New York City in which surplus energy and by-products are exchanged between a sewage treatment works and a nearby total energy plant which provides a nearby housing development with all its electricity, hot water, heating and air-conditioning. Sewage effluent can be used as cooling water in the electricity generation plant (18).

CONCLUSION

37. While there is fairly extensive use of sewage effluent by industry world wide, it is usually restricted to low grade uses such as cooling water and water for washing or quenching. Only in areas of very limited water resources is effluent re-used for high grade uses, such as boiler feed water, where treatment costs are high.

38. The potential operational problems associated with the composition of sewage effluent, principally in recirculating cooling systems, are fairly well characterised and can be overcome by careful control. As a result effluent is suitable for many industrial uses. Its use will, however, generally be restricted to heavy industries such as basic metals processing, with large water demands for low-grade uses. The newer high technology industries such as electronics, while using significant volumes of water have very stringent water quality requirements, and even mains water required a considerable degree of treatment to render it suitable for process use.

REFERENCES

1.   MINISTRY OF HOUSING AND LOCAL GOVERNMENT. Taken for Granted.   Report of the Working Party on Sewage Disposal. HMSO, London, 1970.

2.   OFFICE OF WATER RESOURCE AND TECHNOLOGY. Industrial wastewater   re-use,   cost   analysis   and   pricing strategies.   US   Department   of   the   Interior, OWRT/RU-80/7. 1981.

3.   HART, O. A case for the use of renovated wastewater by industry.   Effluent and Water Treatment Journal. 1979, 19, No. 11, 563.

4.   NEYSADURAI, A.   Wastewater Re-use - the Singapore Experience.   Seminar   on   Pollution   Control   and Management. Singapore, 1977.

5.   WELLS, D.M., SWEAZEY, R.M. and WHETSTONE, G.A. Long term experiences with effluent re-use.   Journal of Water Pollution Control Federation. 1979, 51, No. 11, 2641.

6.   MINISTRY OF THE ENVIRONMENT. Regulations for the use of   industrial   water.   Sewerage   Procedures   & Requirements for Planning Approval, Building Plan Approval & Sewerage Plan Approval. Singapore. 4th Edit. April 1981.

7.   CULP, WESNER, CULP and HUGHES.   Water re-use and recycling.   Volume 1 Evaluation of needs and potential. US Department of the Interior, Office of Water Research and Technology, OWRT/RU-79/1. 1979.

8.   COX, G and HUMPHRIS, T. The use and re-use of sewage effluent.   Proceedings of a symposium, London, April 1976. Cooling Water Association London. 1976.

9.   HORSEFIELD, D.R. and GOFF, J.D. Wastewater resources in north central Texas. Journal of American Water Works Association. 1976, No. 7, 357.

10.   CAIN, C.B., KUESENER, J.W. and LAZARUS, E. Design of 90 mgd wastewater reclamation plant. Journal of Environmental Engineering Division, ASCE. 1981, 107, No. EEI, 29.

11.   DEAN, R.B. and LUND, E. Water re-use. Problems and Solutions. Academic Press, London, 1981.

12.   EYNON, D. Wastewater treatment and re-use of treated sewage as an industrial water supply. Chemical Engineer. London, 1970, No. 235, CE6-CE7 and CE13.

13.   KALINSKE, A.A. Wastewater re-use in Saudi Arabia : the new oasis. Water and Wastes Eng. 1980, Vol. 17, No. 6, 28.

14.   RODDY, C.P. et al. Sewage water effluent will cool new power plant. Journal Power Eng. 1980, Vol 84, No. 4, 96.

15.   MARGLER, L.W., ROGOZEN, M.B. Effects of coal slurry on wastewater bacteria and bacteriophage. Journal of Water Pollution Control Federation. 1980, Vol 52, No.1, 53.

16.   HARKER, R.P. Recycling sewage water for scouring and dyeing. American Dyestuff Reporter. 1980, Vol 69, 1, 28.

17.   ALTOWSKI, G.S. Use of purified wastewater in Moscow's industrial water supply. Wasserwirtschaft Wassertechnik. 1979, 29, No. 8, 278.

18.   STARKEY, H.G. and FORNDRAN, A. Starrett City/New York City energy exchange project. Journal of Water Pollution Control Federation. 1983, No. 7, 1941.

Discussion on Papers 4 and 5

MR R. MARTINDALE, CBI, London, UK
To industrialists, leaving aside manufacturers of water and
sewage treatment equipment, water is a mere incidental to
their main objective of providing the goods or services to
their customers. They are looking for secure, cheap and
trouble-free supplies - trouble-free being identified with
good quality water. The basic disposition is for good quality
water - used water is very much a second choice.

Professor Okun and the other speakers have identified that
it is only when water is scarce or expensive - and usually the
two go together - there is a stimulus to move to lower quality
or at least to look more closely at it. This does not,
normally, describe the situation in the UK where water is
abundant and, by comparison with elsewhere, cheap.

However, the cost of supply is only one aspect of the
subject - effluent disposal is the other and this can
constitute a sizeable and necessary expense for an
industrialist. What Mr Squires shows in Paper 4 is how this
problem may be resolved with benefits to the bill for water.
Dr Taylor and Miss Denner point out in Paper 5 that the use of
sewage effluent by industry is usually restricted to low grade
uses. My view is that it is likely to stay that way, the main
reason being objections against use for production purposes.
Aesthetically it is not a good selling point for many
products. There is also technical uncertainty about the
quality of the supply - one does not prejudice an expensive
product for the sake of a marginal saving, in total cost
terms, by using low grade water.

What exactly are the risks associated with dual supplies?
We are at risk of talking about the subject without pointing
out the problems.

I share the view that industrial reuse has potential
provided that the saving on water costs exceeds the additional
cost of treating to a satisfactory discharge standard. But I
notice that paragraph one of Paper 4 refers to experience with
the imposition of water restrictions. Therefore, I must ask
the author if he could give some guidance on the extra costs
to make water reuseable for process purposes over and above
straight treatment for discharge for some of the industries

with which he has experience. How sensitive are process
economics to having a reliable market and market price for
recovered materials?

PROFESSOR B. DIAMANT, Ahmadu Bello University, Zaria, Nigeria
Mr Squires did not emphasize sufficiently the ultimate target
of industrial wastewater reuse - the reduction of the
biochemical strength of the final effluent, thus leading to
stream pollution control.

In developing countries, development of industry is
considered to be the most applicable answer to the chronic
unemployment problems. However, most developing countries do
not yet have proper industrial waste disposal legislation.
The reuse and production of by-products from waste can be an
incentive for industry to effect reuse and thus to achieve
some level of purification of the wastewater.

MR J. A. CROCKETT, Gutteridge, Haskins & Davey Pty Ltd,
Melbourne, Australia
Paper 4 does not include any information on the cost of
membrane filtration in the applications covered. Would the
process be economic in any of the cases cited at this stage or
will it become economic with the introduction of new
membranes?

Table 3 of Paper 5 sets out apparently rigid quality
guidelines for various water uses in industry including makeup
to recirculated cooling systems.

Such rigid guidelines, at least in the case of cooling water
makeup to recirculated evaporative cooling systems, have
little basis and tend to discourage reuse. Flexible
guidelines for the quality of the water within the system are
available (for example reference 8 in the paper). Such
guidelines are the only ones with a sound basis and have to be
applied, taking into account the concentration ratio and
materials of construction.

In all cases where reuse is being considered the particular
situation must be taken into account. Continued publication
of rigid guidelines as in this paper, as in Appendices 2 and 3
of Paper 7 and elsewhere are a significant discouragement to
reuse.

PROFESSOR D. A. OKUN, University of North Carolina, USA
I am surprised that mention was not made of effluent charges
as a device to encourage industrial recycling. If an industry
is charged for its discharge as a function of both quantity
and strength, industry will be encouraged to conserve both the
water and the waste contained therein. For example, instead
of wash-up with water, the clean-up will be done by vacuum
systems.

DR M. R. G. TAYLOR and MISS J. M. DENNER, Paper 5
In reply to Mr Martindale, we agree that in the UK mains water
is relatively cheap and, if there are no restrictions on

supply, there is little incentive for the reuse of sewage effluent by industry except for low grade uses involving large volumes of water for which mains water would be too expensive. However, sewage effluent, while being of lower quality compared with mains water, can represent a stable source of supply in terms of quantity and even quality, and in areas where water resources are limited it can be the most reliable source of supply. Even in the UK in times of severe drought, such as the summer of 1976, industrial users of sewage effluent would in some areas have found themselves in a much more secure position with regard to water supply than those using wholly mains water.

The main risks associated with dual supplies are the deliberate or accidental use of the wrong grade of water and the contamination of the higher quality supply by the lower grade water as a result of cross connections. This should be largely avoidable by careful design to segregrate the pipework carrying different qualities of water as far as possible and to establish a strict identification code for the supply networks. Dual supply systems are in operation in industrial and residential areas of Singapore using treated sewage effluent, and in Hong Kong and areas of the Middle East using sea or brackish water, apparently without serious problems.

In response to Mr Crockett, the inclusion of Table 3 was simply to give an example of the standards either in force or recommended in practice. We agree that enforcement of rigid guidelines could tend to discourage effluent reuse and that there should be flexibility to reflect the quality of the effluent available and the acceptable range of quality in terms of both the process involved and the materials of construction of the plant. This is in accord with the general policy with regard to effluent standards in the UK, which is based on a 'horses for courses' approach rather than blanket standards for all discharges.

6 Urban effluent reuse for agriculture in arid and semi-arid zones

M. B. PESCOD, OBE, BSc, SM, CEng, FICE, FIPHE, MIWPC, MRSH, Professor of Environmental Engineering and U. ALKA, BSc(Hons), MEng, Department of Civil Engineering, University of Newcastle upon Tyne, UK

SYNOPSIS. The paper reviews the health and agronomic problems associated with the reuse of urban effluent in irrigation. It is suggested that effluent quality parameters of importance to plants and soil structure, as well as those of health concern, should be considered when treatment decisions are being taken. Attention is drawn to the benefits of lime treatment in upgrading saline effluents and a case study is used to illustrate the environmental impact of alternative treatment processes. Multiple criteria analysis is introduced as a useful planning tool which is capable of integrating the wastewater treatment and irrigation components of effluent reuse schemes.

INTRODUCTION

1. Over the past decade, urban water supplies have improved considerably in developing countries and with this advance has come an increasing problem of wastewater disposal. Table 1 shows the projected effluent production in some cities of developing countries for the year 2000 A.D. In arid and semi-arid areas, as well as in many tropical areas, it is imperative that this effluent be reused and agricultural reuse is consistent with the escalating demand for food.

Table 1. Projected effluent production by the year 2000 A.D. in some cities in developing countries

	Total Population (Millions)	Projected Yearly Effluent Production (M m³)
Mexico City - Mexico	31.0	2,602
Sao-Paulo - Brazil	25.8	2,166
Cairo - UAR	13.1	1,100
Karachi - Pakistan	11.8	991
Teheran - Iran	11.3	949
Lagos - Nigeria	6.9	579
Addis Ababa - Ethiopia	5.6	470

2. Effluent reuse is not a new concept; controlled wastewater irrigation has been practised on sewage farms in Europe, America and Australia since the turn of the century and the value of wastewater for crop irrigation is becoming increasingly recognised in arid and semi-arid countries. However, few developing countries have been actively involved in controlled

Reuse of sewage effluent. Thomas Telford Ltd, London, 1984

93

effluent reuse, because urban areas have not been sewered to any great extent. As more sewerage systems are introduced, it is inevitable that effluent reuse in irrigation will be considered.

3.   In many developing countries, governments are now looking for ways of improving and expanding agricultural production and irrigation is an essential component of most developments. Conventional approaches, such as large-scale irrigation, are of great complexity, have a vast appetite for finance and are sensitive to errors in planning. In order to meet the total water needs in arid regions, innovative approaches to water technology must be sought. It is essential that urban effluents be reused in agriculture and their treatment and application in irrigation must be considered as an integrated system.

PROBLEMS OF WASTEWATER REUSE

4.   When sewage effluent is to be used for irrigation both public health and agronomic effects must be considered. Public health considerations are centred around pathogenic organisms that are or could be present in the effluent in great variety. These could produce diseases in farm workers who irrigate and handle the crops, in people who consume the crops and in people who inhale effluent aerosols from spray-irrigated fields. The build-up of toxic materials within the soil, and subsequently in plant and animal tissues, and the danger of them entering the human food chain are possibilities which must be avoided. The leaching of materials such as nitrates and toxic soluble chemicals into groundwater must also be considered. Agronomic concerns are associated with the effects of the inorganic constituents of sewage effluent on the yield of crops and on the structure of the soil, as affected by the accumulation of salts, etc. These problems have been the subject of considerable research interest and some of the findings are described in the following sections.

Public Health Concerns

5.   Risk assessment. The mere presence of an infectious agent in an effluent is not sufficient cause to declare the wastewater unsafe. Hutzler and Boyle (ref. 1) have indicated that 'Even the most dreaded hazard poses no risk if people are not exposed to it'. It is important, therefore, in assessing the health hazards of wastewater reuse to establish the relative importance of various routes of transmission, from direct contact with the wastewater, through food or air, to indirect contact. Important parameters are: the concentration of the infectious agents, the amounts ingested, the duration of exposure and the characteristics of the exposed population. The first three are closely related and are controlled by the survival of pathogenic organisms in the effluent, on crops and in the soil.

6.   Pathogen survival. The concentration of micro-organisms in domestic sewage is dependent upon a number of complex factors, including demography, location and season. The survival of pathogenic organisms in wastewater is also highly variable, as illustrated in Table 2. Whereas organisms such as salmonella

typhosa have relatively short survival times in wastewater, other pathogens, including bacterial spp, ascaris ova and certain enteric viruses, appear to be highly resistant to environmental stress. Although studies (ref. 2) have indicated that microorganisms do not penetrate healthy undamaged surfaces of vegetables exposed to sunlight, pathogens can survive for extended periods inside leafy vegetables or in protected cracks or stems.

Table 2.   Survival of excreted pathogens (at 20-30°C) (ref. 33)

Type of Pathogen	Survival Times in Days			
	In Faeces, Night Soil and Sludge	In Fresh Water and Sewage	In the Soil	On Crops
1. Viruses				
Enteroviruses	<100 (<20)	<120 (<50)	<100 (<20)	<60 (<15)*
2. Bacteria				
Faecal Coliforms	<90 (<50)	<60 (<30)	<70 (<20)	<30 (<15)
Salmonella spp	<60 (<30)	<60 (<30)	<70 (<20)	<30 (<15)
Shigella spp	<30 (<10)	<30 (<10)	-	<10 (<5)
Vibrio cholerae	<30 (<5)	<30 (<10)	<20 (<10)	<5 (<2)
3. Protozoa				
Entamoeba Histo-lytica cysts	<30 (<15)	<30 (<15)	<20 (<10)	<10 (<2)
4. Helminths				
Ascaris Lubricoides eggs	many months	many months	many months	<60 (<30)

*Figure in brackets shows the usual survival time

7.   The density of microorganisms in aerosols is a function of the density of the specific organisms in the effluent, the amount of effluent aerosolised, the effect of shock on the organism and biological decay of the organisms with distance in the downwind direction. A number of studies (ref. 3,4) have attempted to quantify the levels of microorganisms emitted in aerosols generated by effluent spray irrigation and wastewater treatment processes. Teltsch and Katzenelson (ref. 5) reported that enteroviruses were isolated from the aerosols emitted from wastewater spray irrigation. Shuval (ref. 6) estimated that between 0.1 and 1% of the effluent sprayed into the air forms aerosols which are capable of being carried considerable distances by wind. Ledbetter et al. (ref. 7) have shown that wastewater treatment plant workers had a higher incidence of influenza than water treatment plant workers, while Katzenelson et al. (ref. 8) showed that persons living in agricultural communities practising wastewater spray irrigation had a higher incidence of shigellosis, salmonellosis, typhoid fever and infectious hepatitis than those who lived in settlements that did not irrigate with wastewater. However, they did not confirm the mode of transmission (whether aerosols, person-to-person contact with field workers, or other).

8.   A number of factors are known to influence the survival of pathogens and indicator organisms in the soil, indicated in Table 2. These include the level of wastewater treatment previously applied as well as soil moisture, temperature, sunlight, pH, antibiotics, toxic substances, competitive organisms, available nutrient and organic matter. The method and time of application of wastewaters and the soil type will also have an influence.

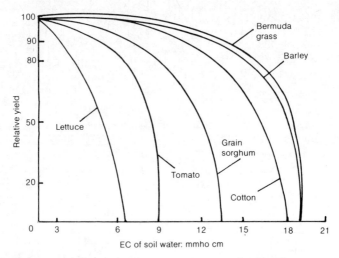

Fig. 1. Salt tolerance of selected crops (ref.35)

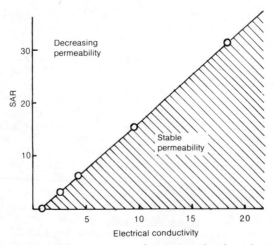

Fig. 2. Threshold concentrations for irrigation water (ref.34)

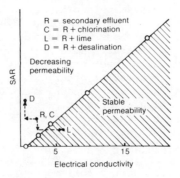

Fig. 3. Effect of alternative upgrading of effluent on soil

9.   Pathogen migration in soil and groundwater contamination.
Contamination of soil and subsequent entry of pathogens into
groundwater is dependent on the survival of organisms during
residence in the soil and the likelihood of their being transported within the soil structure. The principal methods of
pathogen and indicator organism transport in soils include
movement downwards with infiltration water, movement with surface runoff and transport on sediments and waste particles. One
of the most important processes that controls the contamination
of groundwaters, however, is the adsorption or retention of
organisms on soil particles. Another process assisting in the
removal of bacteria and viruses from water percolating through
the soil is filtration. Laboratory and field experiments have
shown that many soils have a high retention capacity for bacteria and viruses (ref. 9,10,11); in general, retention
increases with increases in soil clay content, cation exchange
capacity and specific surface area.

Toxic Metals Accumulation
10.   According to a summary on land disposal of toxic substances and water-related problems (ref. 12), the availability of
heavy metals to plants, their uptake, and their accumulation
depend on a number of soil, plant and other factors:

Soil factors
(a)   Soil pH - toxic metals are more available to plants
      below pH 6.5
(b)   Soil phosphorus - phosphorus interacts with certain
      metal cations to decrease their availability
(c)   Organic matter - organic materials chelate and complex
      heavy metals so that they are less available to plants.
(d)   Cation-exchange capacity (CEC) - This factor is impor-
      tant in the binding of metal cations. Soils with a
      high CEC are safer for effluent irrigation.
(e)   Moisture, temperature and evaporation - these affect
      plant growth and uptake of metals.

Plant factors
(a)   Plant species and varieties - vegetable crops are more
      sensitive to heavy metals than are grasses.
(b)   Organs of the plant - grains and fruits accumulate low-
      er amounts of heavy metals than leafy tissues.
(c)   Plant age and seasonal effects - the older leaves of
      plants will contain higher amounts of metals.

Other factors
Reversion - with time, metals may revert to available forms
in soil and vice versa.
11.   McPherson (ref. 13) noted that although there is a linear relationship between the accumulation of heavy metals (in
both the soil and vegetation) and the total volume of sewage
effluent applied over the years, investigation of the livers
and kidneys of cattle grazed on sewage-irrigated pasture indicated no increase with age. Feigin et al. (ref. 14) reported
that the use of municipal effluent irrigation does not present
a particular hazard in relation to heavy metals. However,
because the mechanism of reversion is not yet understood, it is

97

better to place some restriction on the level to which heavy
metals can be allowed to accumulate in the soil.

12.   In respect of groundwater contamination, Roberts et al.
(ref. 15) reported that trace metals have travel times that are
greatly in excess of the travel times of percolating effluent,
with the effluent travelling nearly 50 times faster. Most
studies (ref. 16,17) have indicated that heavy metals and toxic
organics concentrate within the first 10 cm of the soil where
effluent irrigation is practised. However, continuous dumping
or disposal of contaminated sludge to land, especially on sandy
soils, could lead to pollution of the groundwater system by
metals such as zinc.

13.   Effect of effluent salinity on plants. Dissolved salts
exert an osmotic effect on plant growth. Osmotic pressure is a
colligative property and is, therefore, related to the total
salinity concentration rather than the concentration of individual salt species. An increase in osmotic pressure of the soil
solution increases the amount of energy which the plant must
expend to take up water from the soil. As a result, respiration is increased and the growth and yield of most plants decline
progressively as osmotic pressure increases. Fig. 1 shows the
effect of salinity on the yield of some selected crops.

14.   Although most plants respond to salinity as a function
of the total osmotic potential of soil water, some plants are
susceptible to specific ion toxicities (ref. 18). Many of the
ions which are harmless or even useful at relatively low concentrations may become toxic to plants at high concentrations.
The ions that are most likely to be toxic to plants are chlor-
ides, sodium, bicarbonates, sulphates and boron.

15.   Effect of effluent salinity on the soil. The physical
and mechanical properties of soil, such as dispersion of particles, stability of aggregates, soil structure and permeability,
are very sensitive to the type of exchangeable ions present in
percolating water. The productivity of irrigated land is fundamentally dependent on its internal drainage, which is a function of the soil profile morphology, pore size distribution and
stability of pore structure. The first two factors are of significant importance in considering effluent reuse in agriculture.
No irrigation scheme can succeed unless the soil profile remains
permeable and this depends both on the proportion of exchangeable cations held by the soil that are sodium (that is, the
exchangeable sodium percentage - ESP) and on the concentration
of soluble salts in the percolating water. Quirk and Schofield
(ref. 19) showed that laboratory hydraulic conductivity of a
soil was a function of the relative proportion of exchangeable
sodium (ESP) or the related solution parameter, the sodium
adsorption ratio (SAR). They identified a threshold relationship between the soil ESP or solution SAR and the total salt
concentration of the percolating solution, as shown in Fig. 2,
which distinguishes stable pore structure from unstable structure. Earlier, Richards (ref. 20) had demonstrated the deleterious effects of adsorbed sodium on agricultural soils and Singer
et al. (ref. 21) demonstrated that adsorbed sodium encourages
soil erosion.

16. Water delivery to crops. A problem is likely to arise
in meeting the water needs of crops. Considerable variation
occurs in the production of the effluent as well as in crop
water demand, on both a diurnal and seasonal basis. Thus,
provision must be made for balancing daily variation between
effluent supply and agricultural demand. The problem of the
seasonal water demand pattern for crops, as well as the occasions when no irrigation is possible (i.e. during the rains),
must also be accommodated. Effluent should, therefore, be
stored in a reservoir from which it will be drawn off as needed,
which will remove the problem of the daily variation. However,
for long-term storage, a critical evaluation must be carried out
to analyse possible trade-offs between storage costs (and benefits from subsequent reuse) and alternative effluent disposal
such as groundwater recharge or into surface water.

17.   Various irrigation methods have been used for crop production, including furrow, sprinkler and trickle systems. Furrow irrigation with treated effluent does not differ from
irrigation with water from wells or streams. However, land lev-
elling should be carried out carefully to avoid puddles of
stagnant effluent. Sprinkler irrigation is commonly used in
many countries but when irrigating with treated effluent,
plugging of sprinkler nozzles could be a problem. Precautionary
measures include installation of gravel filters as strainers
and enlargement of the diameter of the nozzles, preferably to
not less than 5 mm. Trickle irrigation seems to be most promising and is now being used in a number of countries, particularly Israel. It has proved to be efficient in arid areas,
because it reduces water consumption. Other irrigation methods
are wasteful of water because the extended area of wetted soil
greatly encourages evaporation. Drip irrigation adopts a
system of pipes placed among the plants or under the soil.
Water carried in the pipes drips onto the soil through outlets
arranged near each plant. Thus, because only a small amount of
soil is watered, evaporation losses are minimised. The rate
and timing of water application are adjusted to minimise runoff
and percolation losses and to enable the smoothing of fluctuations in water availability. Minimising evaporation also
limits soil salination caused by excessive evaporation.

WASTEWATER TREATMENT FOR REUSE IN IRRIGATION
Approach to Effluent Treatment

18.   There are many treatment processes which can be combined
to produce an end-product from urban wastewater that would be
acceptable for uses ranging from grass irrigation to human consumption. The choice of treatment technology is very important
because of the economic consequences of the decision, particularly in poor developing countries. Unnecessarily costly treatment will divert scarce resources away from other developmental
uses. Because wastewater treatment is expensive and because
crops differ very much in the level of effluent treatment they
require, decisions on the choice of crops and level of treatment is crucial to the economic and environmental success of
the system. The present universal trend towards secondary

biological treatment followed by tertiary treatment, usually rapid sand filtration and chlorination, is not a rational approach to the wide diversity of problems raised by effluent reuse in irrigation. It is not logical to assume that an intermediate effluent quality, suitable for discharge to a surface water, is necessary in moving towards a final effluent which must be adapted to the local soil and plant conditions.

Effluent Quality Standards

19.   The prime water-quality objective in any reuse scheme is to prevent the spread of waterborne diseases. For agricultural reuse, protection of public health can be achieved by limiting people's exposure to the reclaimed water and by reducing the concentrations of pathogenic microorganisms in the reclaimed water. Various standards for effluent reuse have been set up and reported (ref. 22,23,24). The Californian standards, reported by Camp, Dresser & McKee (ref. 24), require that reclaimed water for irrigating food crops must be adequately disinfected and filtered, with a median coliform count of no more than 2.2/100 ml. WHO (ref. 22) recommends that crops eaten raw should be irrigated only with biologically treated effluent that has been disinfected to achieve a coliform level of not more than 100/100 ml in 80% of the samples. In setting up such standards, the environmental impact of effluent reuse has been given a secondary position and this has resulted in an almost total neglect in practice. Even when salinity is considered it is normally only from the point of view of crop tolerance and consequent loss of crop yield.

Upgrading for Effluent Reuse

20.   Secondary biological treatment has been found to be inadequate for the removal of pathogenic organisms, heavy metals and salinity, which are the parameters of principal concern in effluent reuse. This has necessitated the upgrading of effluents from these treatment processes by the addition of a tertiary treatment system, commonly rapid-gravity sand filtration and chlorination. Although chlorination of wastewater has long been implicated in the formation of tri-halomethanes, which are carcinogenic, it still remains the most popular method of effluent disinfection, mainly on the grounds of cost. Even if the contribution of dissolved chlorine to the total chloride ion concentration of the effluent is neglected, chlorination has only the effect of eliminating microorganisms.

21.   A method of disinfection currently being investigated is treatment with lime. Most bacteria and viruses die rapidly at pH levels above 11.0 (ref. 25,26) and this pH is easily reached using lime. The lime not only kills but also physically removes bacteria that are attached to solids and thus provides an additional mechanism for pathogen reduction. However, lime has other effects, including its ability to precipitate most heavy metals in solution. Another benefit of lime treatment is the increase in electrical conductivity of the effluent at the same time as a lowering of SAR. In effluent upgrading, salinity removal has always presented a problem. Kasper and Ellis

(ref. 27) reported that the desalination process most commonly used in municipal wastewater reclamation in the U.S. is reverse osmosis (RO) but this is an expensive and complex process. In addition, Crossley (ref. 28) has reported that the rejection of different ions (by RO) depends on their valency. Divalent ions are rejected more effectively than monovalent ions and a membrane which rejects 93% of Na^+ or $C\ell^-$ will reject 98% of Ca^{2+} or SO_4^{2-}. Thus the effect on wastewater effluent will generally be to lower the electrical conductivity but raise the SAR.

PLANNING FOR REUSE
Welfare Economics
22.   The accepted theory of welfare economics states that optimal allocation of resources results when the marginal total cost of providing goods or services is equal to the marginal total benefit of receiving the goods or service. Thus the economist's goal for effluent reuse would be to have it produce quantities of goods that will equate marginal total costs and marginal total benefits. However, in assessing the desirability of effluent reuse, the significance of economic returns must not be allowed to overshadow environmental consequences. In classifying potential damage there are certain features which must be analysed. First, pollutants such as pathogens and toxic chemicals might have immediate effects, with obvious social damage, while others, like salinity, are slow acting. Social damage from this latter type may be considered to proceed in a stepwise fashion such that each increment might be judged harmless until some threshold is exceeded, and effects tend to be invisible up to the threshold. Thus the typical response to high effluent salinity has been to ignore it and its negligible effect up to the damage threshold has militated against avertive measures being taken.

23.   It is essential that social decisions are based not merely on the present situation. Decision-makers must consider how the welfare of future generations will be affected by decisions taken now. A current generation that cares little for future generations can be thought of as weighting the early costs and benefits more than the later ones. This is essentially the situation with effluent reuse as it is practised now.

Case Study
24.   The effluent reuse project at Al Ain in the United Arab Emirates (ref. 29), can be used to illustrate the point made in the previous section. In that case, the wastewater was described as having an average TDS of 1500 mg/l (2.3 mmhos/cm) with a sodium adsorption ratio (SAR) of 7.7. The topography and climate of Al Ain is: 'typical of an arid classification.....; Rainfall of the order of 60 mm ....; the effect of high temperature, low humidity (annual average 45%) and steady light winds combine to produce evaporation rates with an average of 3900 mm'. Effluent reuse is currently planned for 'roundabout and roadside irrigation' but 'future extension to agricultural and horticultural irrigation' is envisaged. Treatment comprised the extended aeration process for biological treatment, final polishing of the effluent by dual-media gravity sand filters and effluent

101

Table 3. Cost-benefit analysis of effluent reuse with different systems (Cost of soil reclamation not included)

Effluent Flow	1892 m³/day (0.5 MGD)			3785 m³/day (1 MGD)			18,925 m³/day (5 MGD)		
Type of upgrading / Crops Grown & returns (£)	Chlorination	Lime Addition	Reverse Osmosis	Chlorination	Lime Addition	Reverse Osmosis	Chlorination	Lime Addition	Reverse Osmosis
Tomatoes (92 days)	23,000	18,400	23,000	46,000	36,800	46,000	230,000	184,000	230,000
Barley (92 days)	2,760	2,760	2,760	5,520	5,520	5,520	27,600	27,600	27,600
Forage Grass (181 days)	8,550	8,550	8,550	17,110	17,110	17,110	85,540	85,540	85,540
Total Annual Benefits	34,310	29,710	34,310	68,630	59,430	68,630	343,140	297,140	343,140
*Total Annual Costs (upgrading only)	9,500	8,750	215,000	17,500	17,000	385,000	54,000	62,500	1,615,000
Nett Annual Benefits	24,810	20,960	-180,690	51,130	42,430	-316,370	289,140	234,640	-1,271,860

NOTES
(1) Capital costs (ref. 23) amotised in 20 years. No interest charges
(2) Cropping pattern adopted to ensure total use of effluent
(3) Costs for irrigation system (i.e. channels for distribution, labour etc.) subtracted from total crop yields, and are assumed to be the same for all upgrading systems
(4) It is assumed that tomatoes will have 80% yield with effluent from liming process (figure 1)

Table 4. Cost-benefit analysis of effluent reuse with different upgrading systems (Incorporating soil reclamation costs)

Effluent Flow	1892 m³/day (0.5 MGD)			3785 m³/day (1 MGD)			18,925 m³/day (5 MGD)		
Type of Upgrading	Chlorination	Lime Addition	Reverse Osmosis	Chlorination	Lime Addition	Reverse Osmosis	Chlorination	Lime Addition	Reverse Osmosis
Annual Benefits	34,310	29,710	34,310	68,630	59,430	68,630	343,140	297,140	343,140
Annual Costs	16,867	8,750	413,333	37,167	17,000	753,333	122,000	62,500	2,833,330
Annual Nett Benefits	17,443	20,960	-379,023	30,943	42,430	-684,703	221,140	234,640	-2,490,190

NOTES
(1) It is assumed that after three years of operation, soil reclamation will be introduced
(2) INCREASED COSTS due to leaching operation and gypsum application for chlorinated effluent and effluent from Reverse Osmosis process.

disinfection with chlorine, selected on the basis of being the
lowest cost solution.

25.   Fig. 1 indicates that a wastewater with an electrical
conductivity of 2.3 mmhos will allow irrigation of any of the
crops, with almost 100% yield, but Fig. 2 suggests that SAR 7.7
in conjunction with 2.3 electrical conductivity will make the
effluent hazardous to soils. Fig. 3 illustrates the effects of
three alternative upgrading methods (chlorination, reverse
osmosis and lime treatment) on the same effluent as it affects
soil stability. Wastewater effluent upgrading with chlorination or desalination, or both, will not affect crop yield but,
singly or in combination, they are likely to lead to harmful
effects on the soil. The effect of liming is the opposite;
this form of treatment will increase effluent conductivity,
which will lead to a reduction in crop yield or even elimination of some of the crops from consideration, but will lead to
stable soil conditions, due to its effect on the sodium adsorption ratio. Table 3 shows the nett benefits from upgrading
effluent using the three methods, where no long-term soil damage
is taken into account. Table 4 shows the nett-benefits when
long-term soil damage is taken into account and demonstrates the
inadequacy and inadvisability of basing effluent reuse decisions
on current monetary values.

Decision Making in Effluent Reuse

26.   Plan evaluation models have become meaningful as
operational tools in policy analysis and the methodology of
evaluation has been developing over the past decade. In the
sixties, cost benefit and cost-effectiveness analysis gained a
great deal of popularity but the difficulty of assigning justifiable price tags to various impacts of policy decisions led to
modified evaluation methods. Multiple-criteria analysis has
more recently made a substantial contribution to achieving an
appropriate evaluation methodology, especially because this
approach did not need to make the assumptions underlying a
price-based evaluation. One of the evident strong points of
this method of analysis is the fact that intangible aspects of
decision making (such as environmental decay, technical complexity, skilled manpower requirement, etc.) could be taken
into consideration. Thus the main aim of the method is to
provide a rational basis for solving problems characterised by
multiple evaluation criteria.

27.   Evaluation essentially lists the possible courses of
action and the assessment criteria, measures the performance
of each option against each criterion and assesses the preferred alternative. For the three upgrading methods already
mentioned, an impact matrix could be set up as follows:

$$
\begin{array}{c c c c c}
 & A & B & C & D \\
\text{Chlorine} & \begin{bmatrix} 3 & 2 & 2 & 2 \\ \text{Lime} & 2 & 1 & 3 & 3 \\ \text{RO} & 1 & 3 & 1 & 1 \end{bmatrix}
\end{array}
$$

A = Cost
B = Effluent salinity
C = Ease of operation
D = Energy requirement

where, the preferred alternative has the highest value. The
many variants of evaluation differ in terms of their number of

103

criteria, the order of steps in the evaluation sequence and in the assessment procedure (ref. 30,31,32).

CONCLUSIONS

28. By incorporating site specific data into a rational evaluation methodology, intuitive decision-making based on convention and precedence in effluent reuse can be superseded by a more objective comparison of feasible alternatives in meeting a range of criteria. Much more information on the performance of effluent treatment processes under different climatic conditions is needed to support such an approach to evaluation of alternatives. Conventional and unconventional sewage treatment processes must be assessed using the effluent-quality parameter of importance to irrigation reuse. Optimization of a combination of unit processes will depend on the overall objectives of treatment, not on the attainment of a preconceived intermediate and irrelevant effluent quality. Treatment objectives must take into account the long-term effects of applying the effluent to the soil in the production of a range of desirable crops.

REFERENCES
1.   HUTZLER N.G. and BOYLE W.C. Wastewater risk assessment. Journal of the Environmental Engineering Division, American Society of Civil Engineers, 1980, 106, 919-933.
2.   LARKIN E.P., TIENNEY J.T. and SULLIVAN R. Persistence of virus on sewage irrigated vegetables. Journal of Environmental Engineering Division. American Society of Civil Engineers, 1976, 102, 29-35.
3.   KATZENELSON E. and TELTSCH B. Dispersion of enteric bacteria by spray irrigation. Journal of Water Pollution Control Federation, 1976, 48, 710-716.
4.   HICKLEY J.L.S. and REIST P.C. Health significance of airborne microorganisms from wastewater treatment processes Part I, Summary of investigation. Journal Water Pollution Control Federation, 1975, 47, 2741-2757.
5.   TELTSCH B. and KATZENELSON E. Airborne enteric bacteria and viruses from spray irrigation with wastewater. Applied Environmental Microbiology, 1978, 35, 290-96.
6.   SHUVAL H.I. Health considerations in water renovation and reuse in Shuval H.I. (ed). Water Renovation and Reuse, Academy Press Inc., 1977, 33-72.
7.   LEDBETTER J.O., HANK L.M. and REYNOLDS R. Health hazards from wastewater treatment processes. Environmental letters, 1973, 4, 225-232.
8.   KATZENELSON E., BUIUM I. and SHUVAL H.I. Risk of communicable diseases infection associated with wastewater irrigation in agricultural settlement, Science, 1976, 194, 944-946.
9.   DREWEY W.A. and ELIASSEN R. Virus movement in groundwater. Journal Water Pollution Control Federation, 1968, 40, 257-271.
10. GERBA C.P., WALLIS C. and MELNICK J.L. Fate of wastewater bacteria and viruses in soil. Journal of Irrigation and Drainage Division, American Society of Civil Engineers, 1975, 157-174.
11. BURGE W.D. and ENKIRI N.K. Virus adsorption by five soils. Journal of Environmental Quality, 1978, 7, 73-76.

12. EPSTEIN E. and CHANEY R.L. Land disposal of toxic substances and water-related problems. Journal Water Pollution Control Federation, 1978, 50, 2037-2042.

13. McPHERSON J.B. Land treatment of wastewaters at Werribee, past, present and future. Progress in Water Technology, 1979, 11, 15-32.

14. FEIGIN A, BIELOROI H., SHALHERET J., KIPNIS T. and DAG J. The effectiveness of some crops in removing minerals from soils irrigated with sewage effluent. Progress in Water Technology, 1979, 11, 151-162.

15. ROBERTS P.V., McCARTY P.L. and ROMAN W.M. Direct injection of reclaimed water into an aquifer. Journal of the Environmental Engineering Division. American Society of Civil Engineers, 1979, 104, 933-949.

16. CHANG A.C., WARNEKE J.E., PAGE A.L. and LUND L.J. Accumulation of heavy metals in sewage sludge treated soils. Journal of Environmental Quality, 1984, 13, 87-91.

17. VALDARES J.M.A.S., GAL M., MINGELGRIN U. and PAGE A.L. Some heavy metals in soil treated with sewage sludge, their effects on yield, and their uptake by plants. Journal of Environmental Quality, 1983, 12, 49-57.

18. MAAS E.V. and HOFFMAN G.K. Crop salt tolerance-current assessment. Journal of Irrigation and Drainage Division, A.S.C.E., 1977, 103, 115-134.

19. QUIRK J.P. and SCHOFIELD R.V. The effect of electrolyte concentration on soil permeability. Journal of Soil Science, 1955, 6, 163-178.

20. RICHARDS L.A. (ed) Diagnosis and improvement of saline and alkaline soils. United States Department of Agriculture Handbook 60, 1954.

21. SINGER M.J., JANITSKY P. and BLACKARD J. The influence of exchangeable sodium percentage on soil erodibility. Soil Science Society of America Journal, 1982, 46, 117-121.

22. WHO. Health aspects of treated sewage reuse. Euro Reports and Studies 42, 1980.

23. COUNCIL FOR SCIENTIFIC AND INDUSTRIAL RESEARCH. Manual for Water Renovation and Reclamation. CSIR Technical Guide K42, South Africa, 1981.

24. CAMP DRESSER & McKEE, Guidelines for water reuse. United States Environmental Protection Agency, Contract No. 68-03-2686, 1980.

25. POLPRASERT C. and VALENCIA L.G. The inactivation of faecal coliforms and ascaris ova in faeces by lime. Water Research, 1981, 15, 31-36.

26. GRABOW W.O.K., MIDDENDORFF I.G. and BASSON N.C. Role of lime treatment in the removal of bacteria, enteric viruses and coliphages in a wastewater reclamation plant. Applied and Environmental Microbiology, 1978, 35, 663-669.

27. KASPER D.R. and ELLIS S.H. Desalination technology for treatment of wastewaters for reuse. Proceedings, Water Reuse Symposium II, Washington DC, Vol. II, 1981, 1342-1362.

28. CROSSLEY I.A. Desalination by reverse osmosis. In Porteous A (ed). Desalination Technology. Developments and Practice. Applied Science Publishers, London, 1983, 205-248.

29.   DEANE A.N., TAYLOR M.R.G. and JONES W.F. Sewage treatment and effluent reuse in arid regions. A case study, Al Ain, United Arab Emirates. Balfours Consulting Engineers internal communication, 1983.

30.   WILSON D.C. Waste management. Planning, Evaluation, Technologies, Clarendon Press, Oxford, 1981.

31.   GOICOECHEA A., HANSEN D.R. and DUCKSTEIN L. Multiobjective Decision Analysis with Engineering and Business Applications. John Wiley & Sons, New York, 1982.

32.   HINLOOPEN E., NIJKAMP P. and RIETVELD P. Qualitative discrete multiple criteria choice models in regional planning. Regional Science and Urban Economics, 1983, 13, 77-102.

33.   FEACHEM R.G., BRADLEY D.J., GARELICK H. and MARA D.D. Sanitation and Disease. Health Aspects of Excreta and Wastewater Management. John Wiley & Sons, Chichester, 1983.

34.   QUIRK J.P. Chemistry of saline soils and their physical properties in Talsma T. and Philip J.R. (eds). Salinity and Water Use, Macmillan Press, London, 1971.

35.   MOORE C.V. Economic evaluation of irrigation with saline water within the framework of a farm, Methodology and Empirical Findings: A case study of Imperial Valley, California, in Yaron D. (ed). Salinity in Irrigation and Water Resources, Marcel Dekker Inc. New York, 1981.

7 Reuse of effluent for agriculture in the Middle East

J. P. COWAN, BSc, CEng, FICE, FIWES, FIPHE, MIWPC, MConsE and
P. R. JOHNSON, DIC, MICE, MIWES, John Taylor & Sons, UK

SYNOPSIS. In the arid countries of the Middle East water
has always been regarded as a valuable commodity. As development in the region has taken off in the last decade it has
almost invariably included ambitious plans for effluent
utilisation for both economic and pragmatic reasons. Many of
the projects commenced in the early 1970s have now reached
operational status and are enhancing the local environment.
In practice it is difficult to precisely differentiate
between schemes of amenity benefit and those for agriculture
in this region. One of the largest projects with which the
Author's firm is involved is for the Government of Kuwait
where the first stage of a comprehensive scheme for effluent
reuse for agriculture and a forest station has recently been
completed. An outline of the scheme is given and other
projects in the area are described and set in the context
of long term national plans for countries in the Middle East.

INTRODUCTION
1.   Throughout the history of the Middle East, life has
been dominated by the vastness and inhospitality of the desert
and the contrasting luxury of fertile areas created where
water could be found. For centuries civilisations have grown,
flourished and declined in the vicinity of the great rivers,
the Nile, Tigris and Euphrates. In the recent decades
advances in technology have raised hopes that the ageless
constraints of the region could be lifted and we have witnessed
the growth of new cities and societies to levels of sophistication which would not have been conceived even 50 years ago.
2.   The growth of new urban communities with their demands
for modern services of every type in such a hostile climatic
environment has created an awareness of the need to deploy
every resource to its ultimate extent. This has been particularly illustrated in the context of water resources where the
installation of piped water supplies and sewerage systems has
created a supplementary resource - wastewater, offering pros-
pects of re-use if treated sufficiently.
3.   Ambitious plans have been formulated by Governments in
the Middle East to utilise this valuable asset which are now
reaching fruition with startling visual results and significance for the future. It should be emphasised that the

Table 1 - Water-Borne Pathogens and their Effect on Health		
Group	Genus	Effects on human health
Bacteria	Salmonella	Typhoid fever, paratyphoid fever, enteritis, salmonellosis, food poisoning
	Shigella	Dysentery
	Escherichia	Enteritis (pathogenic strains)
	Vibrio	Chlorera, enteritis, food poisoning
	Clostridium	Gas gangerene, tetanus, botulism, food poisoning
	Leptospira	Leptospirosis
	Mycobacterium	Tuberculosis, skin granuloma
Viruses	Poliovirus	(Fever, poliomyelitis, enter-(itis
	Coxsackievirus A)	(Headache, muscular pain
	Coxsackievirus B)	(Nausea, meningitis
	Echovirus)	(Diarrhoea, hepatitis
	Adenovirus	Fever, respiratory infection, enteritis, inflammation of the eyes (conjunctivitis), involvement of central nervous system
	Reovirus	Common cold, respiratory tract infections, diarrhoea, hepatitis
	Hepatitis A virus	Infectious hepatitis (fever, nausea, jaundice)
Protozoa	Entamoeba	Amoebic dysentery
	Giardia	Giardiasis
Helminths Trematodes	Schistosoma	Schistosomiasis (Bilharzia)
Cestodes	Taenia	Tapeworm infestation in man, in cattle eating eggs (T. Saginata or T. Solium respectively), eggs develop into the Cysticerus stage
Nematodes	Ascaris	Roundworm infestation
	Anchylostomum	Hook worm infestation
	Heterodera	Potato cyst eelworm

Notes
*Not all species within a genus and not all types within a
 species need be pathogenic
+Not all symptoms are produced by one species and not all
 symptoms may be present at the same time.

Source: UK Water Research Centre - 1978

objectives of sewage effluent re-use differ significantly in different countries and involve the full spectrum of measures from low and intermediate technology to intensive and sophist- icated treatment to product water which conforms with inter- national standards for drinking water.

4.   The authors' firm has been privileged to have been involved in many schemes of public health engineering in the Middle East, some of which date back to the early years of the twentieth century.   All the more recent projects now involve, to a greater or lesser degree, the planned utilis- ation of sewage effluent.   To a large extent, the re-use philosophy has advanced further than the development of experiential knowledge, giving rise to fine judgements as to the risks implicit in the implementation of schemes.   In addition, contrasts in social and religious attitudes have further complicated the establishment of acceptable standards on an international basis.

5.   In these circumstances we have taken the view that the health of the community is paramount and, for this reason, subsequent sections of the paper place some emphasis on health.

PUBLIC HEALTH CONSIDERATIONS
6.   The transmittal of waterborne diseases is well documented and understood particularly in those countries of the developing world that have direct experience of them. Every effort is taken to safeguard potable supplies but the transmission of disease by treated effluent is not so well understood.   It is even more essential that the same degree of caution is applied to effluent utilisation projects.   A wide range of diseases can be transmitted by means of polluted water, or are inadequately treated effluent.   These can be divided into three main groups as below:-

Micro-Biological Contamination
7.   The major health hazard presented by domestic waste- water is due to the pathogens it contains.   Pathogens are defined as agents which cause disease in man, animals and plants.   The crude domestic sewage of a community carries the full spectrum of pathogenic organisms associated with the enteric diseases endemic in the community.   Brief descriptions of the most important micro-biological pathogens follow and are summarised in Table 1.

8.   Bacteria.   The most important diseases caused by bacteria and conveyed by effluent and sludge are typhoid, paratyphoic, cholera, bacillary dysentery, enteritis, salmon- ella poisoning.   The bacteria are much reduced by wastewater treatment but are still present after biological treatment and preclude the use of effluent in this form for uncontrolled purposes.   Their inability to survive and multiply in the environment of a sewage treatment works is the principal reason for their reduction in numbers.   Adequate chlorination is highly effective in the elimination of bacteria from

effluent, as also is lagoon treatment where the incidence of sunlight is a major contributory factor.

9. Viruses. Information on the transmission of viral disease through wastewater is limited but pathogenic viruses are normally present. Diseases which could be transmitted by this method include infectious hepatitis, poliomyelitis, enteric diseases and some respiratory and eye diseases. In general the viral content of wastewater is much less than the bacterial content but viruses are more resistant to treatment processes and a lesser degree of removal is achieved, even with chlorination. It is believed, however, that viruses cannot survive for more than approximately 7 days in effluent.

10. Protozoa. Untreated sewage can be expected to carry cysts of the more common protozoa such as entamoeba histolytica which causes amoebiasis and amoebic dysentery. The protozoan cysts which cause the disease to be transmitted are unlikely to be present after tertiary treatment filtration, but will remain in the sewage sludge.

11. Helminths. These are worms or flukes causing debilitating diseases. Numerous worms which are parasitic in man and animals can be present in wastewater. The mechanisms of transmission of the diseases is through the eggs, which can survive for long periods. They are deposited on soil or crops and picked up by physical contact. Typical of the diseases thus caused are tapeworm and hook worm infections, schistosomiasis and ascariasis. Ascaris is an important consideration and a variety, taenai saginata, which is not a common human infection, could have a serious effect on cattle fed on fodder crops irrigated with effluent treated only to secondary standards. The eggs of the worms are resistant to first stage treatment processes, but can be removed from the effluent by a combination of sedimentation and tertiary treatment. However, eggs of ascaris would still be present in sludge after heated digestion, and can only be killed by heating to about 55°C for two hours, or more slowly, by dessication or prolonged exposure in direct sunlight.

Chemical Contamination

12. A second group of diseases can arise from the constituents of wastewater and require consideration. Apart from common chemicals such as lead, nitrogen and sodium which, in excessive quantities, can have significant effects on health, there are nowadays an ever increasing number of complex organic chemical compounds which are suspected of being carcinogenic. Where discharges from industry contribute to the wastewater being treated, an even wider range of chemicals may appear in the effluent. If irrigation is carried out with such effluent, the contaminant may enter the food chain with potentially harmful effects. Of particular concern are the contaminants which may have adverse effects on public health in the long-term even if present only at trace levels. The main chemical contaminants which may be present in wastewater together with the diseases associated with them

110

are detailed in Table 2. In comparison with the health
problems associated with biological contaminants, the health
problems associated with chemical contaminants are less
significant, since most sewage in the Middle East, where
there is little heavy industry, is derived from domestic
sources.

Table 2. Contaminants in wastewater and associated diseases

Chemical Contaminant	Disease
Lead	Lead poisoning
Nitrate	Malhaemoglobinaemia
Sodium	Hypernatratmia
Organic halogens	Cancer
Polynuclear aromatic hydrocarbons	Cancer

Aquatic Insects
    13.   Aquatic insects can also present risks to public
health. The most important of these are the mosquito and the
tsetse fly which carry malaria and sleeping sickness respectively. However, these diseases are not endemic throughout
the Middle East but any large open bodies of wastewater such
as oxidation or maturation ponds should be monitored at
regular intervals to check on the incidence of mosquito larvae.

HEALTH HAZARDS OF WASTEWATER UTILISATION
    14.   The transfer of pathogenic organisms or other contam-
inants of public health significance from wastewater to man
may be affected by one or more of the following methods:-
(a)   Consumption of contaminated food or water
(b)   Insect bites
(c)   Breathing of air contaminated with fine droplets of
       polluted water
(d)   Body contact with contaminated wastewater or soil
Of these the first is the most important as far as the general
population is concerned, whilst the second method is, from
experience to date, not a significant problem. The latter
methods listed affect only those whose work or leisure
activities bring them into direct contact with the effluent.
    15.   Many studies into survival times of pathogenic
organisms in soil have been undertaken mainly in the United
States. The variation in survival times is large as indicated
in Table 3. In general, high soil moisture content, low
temperatures, high soil pH and presence of organic matter
tend to increase survival times; whilst sandy soils, high
temperatures, low soil pH, sunlight and low humidity tend to
reduce them. Thus the conditions in the Middle East generally tend to reduce the survival times. The survival times

111

of pathogenic organisms on wastewater irrigated crops tend
to be lower than in soils as indicated in Table 3.

Table 3.   Pathogen Survival Times in Soil and on Crops

Pathogen	Survival Times	
	In Soil	On Crops Irrigated with Wastewater
Bacteria	Sometimes over a year, but generally less than two months	Up to six months, but generally less than one month
Viruses	Up to six months, but generally less than three months	Up to two months, but generally less than one month
Protozoa	Up to ten days, but generally less than two days	Up to five days, but generally less than two days
Helminths	Up to seven years, but generally less than two years	Up to five months, but generally less than one month

16.   The transfer of wastewater irrigated produce from
field to market and thence into the home presents the greatest risk to the general public. During this period there
may be cross-contamination with other products, contamination arising from handling the produce and contamination of
food preparation environments. Cooking of produce at a
temperature in excess of $75^{o}C$ will kill bacteria, protozoa,
helminths and most viruses. The greatest risk is from
fruit and vegetables that are eaten raw.

17.   The pathogenic organisms having survived and been
ingested may now cause illness. However, the organisms
must be swallowed in sufficient quantities to exceed a
person's threshold of susceptibility and, of course, infants,
elderly persons and malnourished persons are generally most
susceptible. Without extreme caution and scrupulous standards, the gradual spread of effluent re-use will inevitably
cause an increase in enteric disease and in extreme
circumstances, an epidemic situation. Specific precautions
must be taken for those personnel directly associated with
effluent re-use projects and criteria which we have
recommended on all projects are outlined in Appendix 1.

EFFLUENT STANDARDS
18.   For any effluent re-use scheme, it is important that
the most suitable sewage treatment process is chosen.
Sewage treatment plants installed in the Middle East include
percolating filters, stabilisation ponds and activated sludge

and extended aeration plants. With the exception of the
filters, these have satisfactorily met the design standards
when operated effectively.

19.   In broad terms, the organic quality is monitored by
the Biological Oxygen Demand (BOD) and the physical properties by the Suspended Solids (SS). When effluent is diluted
by discharge to a river after treatment the standards for
these two parameters have been traditionally of the order
20 mg/l BOD and 30 mg/l SS. For schemes where no subsequent
dilution is feasible and particularly where effluent re-use
is envisaged, additional (tertiary) treatment is necessary.
Again, by consensus, this standard is typically 10 mg/l BOD
and 10 mg/l SS. Effluent achieving these standards is
generally of sufficiently high quality to prevent a build-up
of solids in the distribution system (causing maintenance
problems) and sufficiently stable and pure to reliably apply
disinfection by chlorination as a primary health safeguard.

20.   The principal chemical characteristics of the
effluent which determine its suitability for irrigational use
are salinity or total concentration of dissolved solids (TDS)
sodium absorption, ratio (SAR), residual sodium carbonate
concentration, substances and metals found in low or trace
concentrations and nutrient concentrations (Nitrogen,
phosphorous and potassium).

21.   Salinity is measured by electrical conductivity (EC).
Irrigation water with an EC of 1650 micro-mhos/cm or greater
is defined as high in salinity. However, such levels are
common in the Middle East, particularly in the Gulf States,
and levels not exceeding 3000 micro-mhos/cm are regarded as
generally acceptable, providing salt tolerant crops are
selected.

22.   High concentrations of sodium ions can cause a
progressive deterioration in soil structure with the de-
floculation of soil particles causing a decrease in the
hydraulic conductivity and aeration and a consequent build-up
of salinity.

23.   Some elements which can cause concern, even in low
concentrations are boron, cadmium, lead nickel, copper, zinc
and selenium. To date concentrations measured have not been
significant but, as the Middle East becomes more industrial-
ised, these elements will need control to ensure that toxic
elements do not adversely influence re-use projects.

24.   The nutrients of most interest are Nitrogen (N),Phos-
phorus (P) and Potassium (K). Excess nitrogen can reduce crop
yields and lead to surplus nitrates being accumulated in
fodder, and thence live-stock, with serious consequences.
In addition, leaching of excess nitrates into groundwater
which is subsequently abstracted for human consumption can
lead to toxicity, particularly in young children (blue baby
syndrome). The likely concentrations of the latter two
elements do not normally give cause for particular concern
but should be monitored to ensure that suitable crops are
selected. Notwithstanding these comments, the nutrient

content of effluent is seen as a valuable economic bonus
arising from effluent re-use in the agricultural context.

INTERNATIONAL STANDARDS

25.   Wastewater, treated or untreated, has been used
in many countries throughout the world for many years. A
report by a World Health Organisation (WHO) panel of experts
in 1973 revealed that standards governing the use of treated
wastewater in agriculture in California, Israel, South
Africa and West Germany exhibited wide divergences and it is
clear that opinions varied on suitable standards. The WHO
Technical Report No. 517 dated 1973 suggested suitable treatment processes to meet the given health criteria for
effluent re-use. These guidelines are reproduced in Appendix
2 and have formed the basis of health criteria for most
authorities for the past decade.

26.   Our recent experience in Abu Dhabi and Kuwait
indicates that the governments of these countries wish to
adopt more stringent health criteria than those suggested by
the WHO.

27.   In Abu Dhabi a new sewage treatment works was
commissioned in 1982. This plant is an activated sludge
extended aeration system which produces an excellent
effluent which is then given further treatment with sand
filters and chlorination. Effluent quality is consistently
better than the 10 mg/1 BOD, 10 mg/1 SS design criteria. All
wastewater is used for irrigation of road central reservations, roundabouts and roadside planting where public contact
is minimal. Parks and gardens where public contact is
common are irrigated with potable water. Effluent quality
is continuously monitored with unsuitable effluent
being diverted to waste (usually due to high salinity).

28.   In Kuwait the first phase of the effluent utilisation project is nearing completion. Wastewater will be
taken from several treatment works, all of which will have
tertiary treatment facilities provided under the project.
Secondary treated wastewater has been used in the past, but
this practice will be phased out. The wastewater will be
used for agriculture and for environmental protection
forestry. The wastewater will only be used for crops which
will normally be cooked prior to being eaten. Forestry areas
will be fenced off to prevent public access.

29.   In Saudi Arabia Draft National Wastewater Regulations (DNWR) were published by the Ministry of Agriculture
and Water in 1982 and are summarised in Appendix 3.
"Environmental Protection Standards" (EPS) which are more
comprehensive and include air pollution have also been issued
by the Meteorology and Environmental Protection Administration and the schemes with which we have been associated are
generally consistent with their requirements. Despite some
apparent anomalies in these standards and in the absence
of international standards, they are welcome steps since the
re-use of effluent is likely to increase significantly in the

Middle East as the traditional water resources become more depleted and less suitable.

SCOPE OF AGRICULTURE

30.   All governments in the region place a high premium on creating flourishing "green areas" within the harsh environment and have readily committed large investments in enhancing the landscape of their cities and adjacent areas. In addition much emphasis has been placed on national plans to produce a significant proportion of the local demand for fruit, vegetables and cattle fodder. In consequence there is little distinction in their approach as to whether specific "green areas" are primarily for commercial agricultural production or amenity. Clearly at the extremes,trees and shrubs planted to enhance the margins of a dual carriageway cannot be classified as agricultural but large scale nursery operations and afforestation projects can be considered as both amenity and the commencement of commercial agriculture.

31.   In this paper we have assumed that the primary purpose of agriculture is to produce feedstuffs for man and animals and secondarily to improve the environment by the development and continued support of afforestation and species of trees, shrubs and flowers tolerant to local conditions. Special cases include planting to stabilise sand dunes and prevent soil erosion by wind or even flash flooding.

32.   Health risks arise principally when crops are grown for animal or human consumption since these affect the population at large, while activities to improve the environment involve lesser risks. However, the technology and precautions to be taken are similar.

Crops for Animal Consumption

33.   These include cereals and alfafa which are grown in large amounts as fodder for dairy animals. Irrigation of these crops is implemented most effectively by sprinkler irrigation. Harvesting, after a period of no irrigation, creates a break in the "health cycle" and handling is largely mechanical thereby reducing risks to farm workers. In addition any remaining bacteria in the effluent applied to the crops is destroyed by the animals' digestive systems and this forms a second barrier to the transmission of disease to the human consumer.

Crops for Human Consumption

34.   These can be divided into two groups - those normally cooked before they are eaten and those which are eaten direct. The former includes potatoes, onions and garlic, for which sprinkler irrigation is a suitable application method. Health risks are low since the drying and storage period which follows harvesting will normally remove most pathogenic risks. The latter group includes tomatoes, aubergines, peppers, citrus fruits, peaches and apricots. These foods

115

are mostly eaten raw and should only be irrigated by a
sophisticated non-contact system such as drip feed, since
this is the only way the health risks can be kept to an
absolute minimum. These methods ensure that water is fed
direct to the plant roots and the produce is thus kept separate from the water supply. Handling of contaminated produce
is thus eliminated.

METHODS OF IRRIGATION
35.   Irrigation has been used by man since the dawn of
civilisation. In many parts of the world tried and tested
methods over the centuries are still used but in more recent
times modern technology has been applied particularly in areas
where wastewater is being re-used. There are three types of
irrigation as follows:
> i)   Flood and Furrow
> ii)   Sprinklers
> iii) "Hi-tech" modern systems

Flood and Furrow
36.   This method accounts for about 95% of the world's
irrigation and is widely used for the irrigation of rice in
the Far East and elsewhere. It was well established in early
civilisations notably in Egypt and Iraq along the river plains
of the Nile, Euphrates and Tigris. Irrigation by flooding is
generally the application of water to basins formed by earth
bunds and fed by subsidiary canals. The water is usually
derived from rivers or canals which have not received any
special treatment. However, in recent times consideration has
been given to sewage effluent being used by discharging to
existing channels or canals in order to supplement existing
supplies. Irrigation by furrow is similar to flood
irrigation except that the basins are structured with ridge
and furrows to give a greater degree of control.
37.   Flood and furrow irrigation dominates the world's
irrigation because it is simple to understand and simple to
implement. It is a low-key but effective technology and
even though its efficiency is relatively low, it is technically eminently suitable for the developing countries if water
is plentiful. However, the cost of providing suitable treatment to effluent will probably limit this method of application for future schemes involving effluent utilisation.

Sprinklers
38.   Sprinkler irrigation has been developed as an effective means of irrigation over the last century mainly as an
offshoot of water distribution technology. Some 3 - 4% of
the world's irrigation uses this method which can be up to
75% efficient. The principal methods of irrigation by sprink-
ling are:-

> i)   Rain Gun - high capacity water jets.
> ii)   Set Pipe System - simply a pipe network system

with numerous rotating nozzles. The pipework can be
rapidly dismantled and re-erected elsewhere.

iii) Side Roll - a system of pipelines supported on large
      wheels in which water is distributed from nozzles
      whilst the side roll equipment is stationary. The
      wheels facilitate movement of the plant either
      manually or by the application of small motors to the
      plant.

iv)   Centre Pivot Sprinkler - a system in which a distrib-
      ution arm fixed at one end, is fitted with several
      support carriages each driven by electric motors,
      enabling continuous rotation of the sprinklers.

Of the above, the latter is the most common form of desert
irrigation but all are suitable for use with treated effluent
provided health controls are given adequate attention.

'Hi-tech' Modern Systems

39. These systems of irrigation are only used for about
1% of the world's irrigation. Whilst the efficiency of
irrigation can be of the order of 90%, the costs of such
systems and more demanding operation and management are the
penalties which must be accepted. The most widely used
modern methods of providing controlled amounts of water to
the crops are trickle strip and drip emission. With trickle
strip irrigation, water is applied directly to the roots of
the plants or row-crops by means of surface mounted small
diameter pipes with small holes bored into the sides of them.
Drip emitters are available as either pressure or non-
pressure compensating types. Spaced in clusters at regular
intervals, they are fed by a system of pipes laid just below
or at ground level. In general two or three emitters are
required per tree or shrub, each giving a discharge in the
order of 4 litre/hr.

TYPICAL SCHEMES

40. In this section selected schemes with which the
authors' firm have been associated are briefly described. A
summary of these and other projects, including types of treatment, effluent standards and re-use application is given in
Appendix 4. Sewage treatment, tertiary treatment, bulk
transfer, storage and distribution facilities which are
essential elements of most of the projects described, are not
covered in detail. In all schemes "appropriate technology"
has been an important consideration.

Kuwait

41. Effluent re-use on a limited scale was established in
Kuwait in the mid 1970s. In the late 1970s we produced a
master plan for the re-use of effluent for agricultural
application and the first phases of the master plan recommendations are now being commissioned. The objective of this
scheme is to make Kuwait self sufficient in several products,
principally in milk, potatoes, onions and garlic by the year

2010.

42.   This scheme includes the bulk transfer of tertiary
treated sewage effluent from three sewage treatment works to
a central distribution and administration centre. From here
it is directed to an existing farm and also to a newly created
agricultural area. Facilities to establish a third area are
included in a further phase of the scheme. The established
farm covers an area of 860 ha, of which only part is under
cultivation. Forage (alfafa) for the dairy industry is the
main crop using side roll sprinkler irrigation but aubergines,
peppers, onions and other crops are grown on an experimental
basis for which semi-portable sprinklers and flood and furrow
techniques are used. The farm is soon to become fully
effective as the first phase of the master plan is commissioned. A newly created area of some 870 ha is scheduled to grow
forage, garlic and onions by side roll sprinklers and will
also produce those experimental crops referred to above on a
commercial basis. As effluent quantities increase, a third
agricultural area of around 1400 ha will be established to
produce similar crops.

43.   Over 10 million m^3/a of effluent is currently being
used which will rise to 125 million m^3/a by the year 2010
constituting one of the most ambitious schemes for agricultural re-use in the Middle East. In addition to the above, an
ambitious programme of afforestation to cover an ultimate area
of 12,000 ha is planned. The prime purpose of the afforestation is to grow a variety of trees for beautification and
wind/ dust breaks particularly for marginal strips along major
highways and to protect new townships. All applications take
special note of possible health risks and the areas are therefore fenced to restrict public access. Additionally farm
workers are to be subjected to regular health checks in order
to monitor the incidence of sickness.

Abu Dhabi
44.   In 1973 we prepared a master plan for the sanitation
of Abu Dhabi which also included recommendations for the reuse of effluent. Detailed design was authorised in 1974 and
construction commenced in 1976. Because Abu Dhabi had no
established agriculture, with the exception of afforestation,
effluent utilisation for agriculture was not considered a
decade ago and the emphasis was on amenity use. However, with
the depletion of water supplies in the surrounding desert,
farmers are moving nearer to the town and a readily available
source of water. There is now scope for the cultivation of
fodder for animals which is currently transported from the
Western Emirates.

45.   The first phase of the scheme has now been substantially completed and includes a treatment works serving a
population of 330,000. It is located 45 km from the town
centre treating some 100,000 m^3/day of effluent to tertiary
standards. This effluent is returned to the city by gravity
where it is distributed by means of pumping stations, ground
storage tanks and pipe networks for amenity use including

central reservations, parks and afforestation schemes. Manual and drip feed application are the principal methods of irrigation.

46. The effect on the environment has been dramatic and now the scheme is well established, it is essential that it is well maintained to prevent any deterioration in the quality of plants and trees which need a constant supply of good quality effluent. As a means to this end, both crude sewage characteristics and effluent quality are closely monitored to ensure that salinity levels (arising from groundwater infiltration) do not increase beyond tolerable levels. In such an event provision has been made for the emergency disposal of unsuitable effluent. In addition to monitoring, various experiments have been carried out including the application of various dyes to effluent to differentiate between potable and effluent distribution networks. This proved to be unsuccessful since plant appetites for the dye were too great. Application of nile blue tracer dye in 2mg/l was technically successful but was finally discontinued because of cost.

Qatar

47. Qatar has a history of effluent reuse dating back for more than a decade and even though on a limited scale, the effect has been significant. Initially effluent from a biological filtration plant was distributed to central reservations, verges and municipal gardens by road tankers and manual hoses. This method has been superseded in recent years as elevated storage tanks and associated distribution mains fitted with simple taps for manual application of effluent have been commissioned. More recently, automatic drip feed irrigation has been introduced.

48. In 1983 a new sewage works incorporating biological treatment by activated sludge and tertiary treatment by rapid gravity sand filters was commissioned with a capacity of 60,000 m^3/day. Of this some 15,000 m^3/day is used for mainly amenity and afforestation, the surplus being pumped to an evaporation lagoon in a remote desert region where it also provides groundwater recharge.

49. The effluent re-use programme is expanding continuously and some 25% of the available effluent is being used for municipal application. Future plans include extension of the irrigation system in Doha to include the race-course. The long term policy envisages all effluent being reused and by the year 2000 it is anticipated 72,000 m^3/day will be available for agriculture, although a firm agricultural policy has yet to be determined. In the smaller towns of Umm Said and Khor, effluent re-use schemes are also being implemented. Effluent from extended aeration plants will be distributed for amenity application via elevated water storage towers, ring mains, and drip feed emitters.

Saudi Arabia

50. Rapid development of cities throughout the Kingdom

has placed a great burden on the Kingdom's natural water
resources and as these have become depleted, effluent reuse has become a topic of some priority. Indeed it is the
publicly stated desire of the Government that all available
treated wastewater in the Kingdom is used in some beneficial
manner.

51.   In the Western Region we have recently designed a
sewage treatment works for Makkah to serve a population of
one million. The works also includes tertiary treatment
facilities (rapid gravity sand filters) followed by chlorination in order that the effluent may be discharged safely
to the wadi, ultimately to be used for agricultural purposes
although the agricultural re-use policies have yet to be
confirmed. The works also include an advanced treatment
plant to treat about 25% of the total flow to potable standards in order that it may supplement domestic supplies,
(toilet flushing, garden irrigation etc.) but this provision
has not been implemented at present. A similar scheme is
currently under construction in Taif, but here too, the final
policy for agricultural application has yet to be formulated.

53.   We are currently finalising the master plan for
effluent re-use in the Qassim Area. This largely agricultural region comprises 1000 sq.km. and includes the major towns
of Buraydah, Unayzah, Al Rass, Al Burkayriyah and Riyad Al
Khabra which have existing or planned sewage treatment
facilities. Whilst volumes of effluent will be small, it is
planned to develop the irrigation of fodder crops on Government controlled farms, landscaping where human contact is
minimal and tree planting for sand stabilisation and environmental protection.

Iraq

54.   The first sewage treatment works was constructed in
Baghdad some 25 years ago and has been extended more recently.
Effluent is discharged into a tributory of the Tigris and,
with ample dilution and further purification in the river,
water is abstracted downstream for the irrigation of crops by
the ridge and furrow method. Whilst water resources have been
sufficient to meet local demand in the past, it is likely that
future pressure on such resources will lead to more intensive
effluent re-use in the future.

Egypt

55.   We are currently engaged in the design of a sanitation system to serve Greater Cairo for a population in excess
of 12m. Two large sewage treatment works are proposed, treating sewage to a standard suitable for discharge to natural
water courses in the Nile Delta where it will receive further
polishing by means of dilution and natural processes.
In such an area the immediate problems are concerned with the
quality of the water abstracted for irrigation purposes and
only secondarily with quantities of timely water supplies.
The construction of new sewerage and treatment facilities will

serve to provide a general improvement in water quality in the Nile downstream of Cairo with a consequent reduction in health risks in irrigated areas.

CONCLUSION

56.   Effluent re-use has reached a level of acceptance in the Middle East which could not have been envisaged twenty years ago. In addition, the commissioning and implementation of sanitation projects over the past few years ensures that the development and refinement of re-use concepts will continue unabated in the foreseeable future.

57.   However, the integration of this final component of local water resources into national plans will give rise to increasing health risks unless a co-ordinated approach to planning and regulation is adopted. It may well prove essential for government agencies involved in the management of different parts of the water cycle to be restructured. In particular there is a need to draft and enforce appropriate legislation covering wastewater discharges, levels of treatment and methods of application of effluent to ensure that re-use projects can be introduced and operated safely within the community.

58.   Whilst legislative provisions are essential, it is no less important that the introduction of re-use concepts is accompanied by widespread public education as to the health risks and necessary precautions. Similarly, the training and education of the necessary managers, technicians and operatives on such projects must emphasise the risks to the health of the community which may arise unless they adhere to the established procedures set up to control these risks.

59.   After a decade of enthusiastic adoption of effluent re-use concepts in the Middle East, many projects are now in use, yielding much satisfaction to the local population and those involved with implementation. The future offers great potential as various master plans approach to completion, providing that the inherent risks to the health of the public are not forgotten.

APPENDIX 1 - CRITERIA FOR THE PROTECTION OF PUBLIC HEALTH

The following criteria is recommended for all re-use projects, including both amenity and agriculture. These adjuncts are essential to the safe use of effluent. However, these guidelines are not always adhered to, and an extensive programme of education in respect of the risks, sources of infection and modes of transmission of potential disease should precede any project.

Effluent Standards:

(i)   Parasite eggs must be effectively removed plus some removal of viruses

(ii) The coliform count must not exceed 100 coliform organisms per 100 ml in 80% of samples of effluent, for food crops for cattle fodder and afforestation.

(iii) The coliform count must not exceed 1000 coliform organisms per 1000 ml in 80% of samples of effluent for food crops for cattle fodder and afforestation

Agricultural Practice:

(i)   Food crops to be eaten cooked or raw: there should be no irrigation for a minimum period of 7 days prior to harvesting

(ii) Pastureland: sheep and cattle must not be permitted to graze on land for a minimum period of 7 days after irrigation

(iii) Sprinkler irrigation: unless special precautions are taken, sprinkler irrigation must not be carried out closer than 100 m to an occupied dwelling or within 50 m of any public road.

Standards of Hygiene for Agricultural Workers:

(i)   Irrigation water must not be used for drinking or domestic purposes.

(ii) Workers must wash their hands in clean water after working in irrigation channels or in any situation where they have been in contact with irrigation water or recently irrigated crops or soil.

(iii) Workers must use adequate protective footwear at all times.

Standards of Hygiene for the Consumer:

(i)   Consistent care is necessary in the preparation of
        raw green vegetables and fruit and adequate standards
        of hygiene for all involved in the handling of food
        would need to be enforced.

   In addition to the following, additional criteria are
applicable to afforestation.

Afforestation Practice:

(i)   Irrigation systems should be designed to minimise
        inadvertent public access by covering such sections
        of open channel which may be adjacent to roads or
        footpaths and similar measures.

(ii)   Adequate warning notices must be displayed indicating
        the use of effluent.

Standards of Hygiene for Forestry Workers:

(i)   Effluent and irrigation water must not be used
        for drinking.

(ii)   Workers must wash their hands in clear water after
        working in irrigation channels or in any situation
        where they have been in contact with irrigation
        water or recently irrigated land.

(iii)   Adequate footwear, such as boots, must be used at all
        times.

Standards of Hygiene for the Public using Afforestation Areas:

(i)   Avoid contact with irrigation water wherever possible.
        In the event of inadvertent contact, wash in clean
        water.

   Technical criteria for this type of effluent use are
basically similar to those for general agricultural use and
the principal concern is to monitor and control the toxicity
which could be caused by trace elements.

	Irrigation			Recreation		Industrial reuse	Municipal reuse	
	Crops not for direct human consumption	Crops eaten cooked; fish culture	Crops eaten raw	No contact	Contact		Non potable	Potable
Health criteria (see below for explanation of symbols)	A + F	B + F or D + F	D + F	B	D + G	C or D	C	E
Primary treatment	●●●	●●●	●●●	●●●	●●●	●●●	●●●	●●
Secondary treatment		●●●	●●●	●●●	●●●	●●●	●●●	●●
Sand filtration or equivalent polishing methods		●	●		●●●	●	●●	●●
Nitrification								●●
Denitrification								●
Chemical clarification						●		●
Carbon adsorption								●●
Ion exchange or other means of removing ions						●		●●
Disinfection		●	●●●	●	●●●	●	●●●	●●● *

Health criteria

A Freedom from gross solids; significant removal of parasite eggs.

B As A, plus significant removal of bacteria.

C As A, plus more effective removal of bacteria, plus some removal of viruses.

D Not more than 100 coliform organisms per 100 ml in 80% of samples.

E No faecal coliform organisms in 100 ml, plus no virus particles in 1 000 ml, plus no toxic effects on man, and other drinking-water criteria.

F No chemicals that lead to undesirable residues in crops or fish.

G No chemicals that lead to irritation of mucous membranes and skin.

In order to meet the given health criteria, processes marked ●●● will be essential. In addition, one or more processes marked ●● will also be essential, and further processes marked ● may sometimes be required.

* Free chlorine after 1 hour.

Source: WHO Technical Report No. 517

APPENDIX 2 - SUGGESTED TREATMENT PROCESSES TO MEET GIVEN HEALTH CRITERIA FOR WASTEWATER UTILIZATION

	IRRIGATION			RECREATION		PONDS AND LAKES				AQUIFER RECHARGE			INDUST-RIAL USE	DISPOSAL TO SURFACE WATER BODY
PARTICULAR CRITERIA	RESTRICTED - LEGUMES, FORAGES & FIELDCROPS OR VEGETABLES MECHANICALLY PROCESSED	UNREST-RICTED	CROPS EATEN RAW, FISH CULTURE	CASUAL HUMAN CONTACT	COMMON HUMAN CONTACT	RESTRICTED	UNREST-RICTED	STOCK WATERING	FORESTRY 25 m OUTSIDE MUNICIPAL LIMITS	OPEN LAND	RESTRICTED ACCESS	DIRECT INJECTION		
	A	B	-	C	D	-	D	D	-	E	-	F	G	H
PRIMARY TREATMENT	**	**	**	**	**	**	**	**	**	**	**		**	
SECONDARY TREATMENT	**	**	**	**	**	**	**	**	**	**	**		**	
(OXIDATION ?)				**	**	**		**						
TERTIARY TREATMENT 1. METHOD UNSPECIFIED	**	**	**		**		**	(**)		**			**	
2. SAND FILTRATION			*											
ADVANCED TREATMENT 1. DENITRIFICATION			*											
2. PHOSPHATE REMOVAL			*											
3. COAGULATION					**		**	**						
4. CARBON ADSORPTION			*			**								
5. DISINFECTION			*		**	**	**	**						

Note - See Sheet 2 for notation

APPENDIX 3 - SUMMARY OF SAUDI ARABIAN MINISTRY OF AGRICULTURE AND WATER (MOAW) DRAFT NATIONAL WASTEWATER REGULATIONS (Sheet 1)

Secondary treatment is defined as follows:-

BOD   - not greater than 20 mg/l (mean for 30 consecutive days)
SS   - not greater than 20 mg/l (mean for 30 consecutive days)
Faecal coliforms - not greater than 100 per 100 ml (mean for 30 consecutive days)
Total coliforms - not greater than 23 per 100 ml (medium no. for 7 days)
pH   - in range 6.0 to 9.0

Tertiary treatment is defined as follows:-

BOD   - not greater than 10 mg/l (mean for 30 consecutive days)
SS   - not greater than 10 mg/l (mean for 30 consecutive days)
Faecal coliforms - not greater than 50 per 100 ml (mean for 30 consecutive days)
Total coliforms - not greater than 2.2 per 100 ml (medium no. for 7 days)

**   Process essential
*   One or more of these processes are required
(**)   Process implied but not stated

PARTICULAR CRITERIA

A - Criteria include warning signs, no animals
    to drink, measures to prevent insect
    breeding, etc.

B - Tertiary treatment implied by MCL specified.

C - Total coliforms not greater than 23 per
    100 ml (last 7 days)

D - Total coliforms greater than 2.2 per
    100 ml (last 7 days).

E - Faecal coliforms not greater than 200 per
    100 ml.

F - Written permit from MOAW who will stipulate
    water quality criteria.

G - Individual users of Ministry of Industry
    may set requirements.

H - Written permit from MOAW who will stipulate
    water quality criteria. Where aquatic life
    is involved criteria are stated in the
    regulations.

APPENDIX 3 - SUMMARY OF SAUDI ARABIAN MINISTRY OF AGRICULTURE AND WATER (MOAW) DRAFT NATIONAL WASTEWATER REGULATIONS   (Sheet 2)

COUNTRY	CITY	SEWAGE TREATMENT			TERTIARY TREATMENT			
		PROCESS	DESIGN CAPACITY m^3/day	EFFLUENT STANDARDS BOD/SS	PROCESS	DESIGN CAPACITY m^3/day	EFFLUENT STANDARDS BOD/SS	REUSE APPLICATION
KUWAIT	Kuwait	AS	342,000	20:30	RGSF&Cl_2	342,000	10:10	A, F, V
U.A.E.	Abu Dhabi	AS	100,000	20:30	RGSF&Cl_2	100,000	10:10	MU, A
QATAR	Doha	AS	60,000	20:30	RGSF&Cl_2	81,000	10:10	MU, A
	Umm Said	EA	4,320	10:10	Cl_2	-	-	MU
	Khor	EA	728	10:10	Cl_2	-	-	MU
SAUDI ARABIA	Makkah		112,000	20:30	RGSF&Cl_2 AT	112,000 90,000	10:10 Potable standards	MU, W, NP W, NP
	Abha	EA	22,500	20:30	RGSF&Cl_2	22,500	10:10	F, A
	Qassim Prov.	EA	75,000	20:30	RGSF&Cl_2	75,000		
OMAN	Khuwair Town	EA	10,000	10:10	Cl_2	-	-	MU, A
LIBYA	Sebha	EA	15,000	20:30	RGSF&Cl_2	15,000	10:10	W, NP
	Yeffren	EA	1,725	20:30	Cl_2	-	10:10	W, NP
EGYPT	Cairo	AS	1,000,000	30:30	Cl_2	-	-	RD, NP
	"	AS	600,000					
IRAQ	Baghdad	AS	68,200	20:30	-	-	-	RD, GA

Notes:

AS = Activated Sludge
EA = Extended Aeration
RGSF = Rapid Gravity Sand Filtration

Cl_2 = Chlorination
AT^2 = Advanced Treatment

MU = Municipal Use
A = Afforestation
F = Fodder

V = Vegetables
W = Discharge to Waste
RD = Discharge to River

GA = General Agriculture
NP = No policy yet determined

APPENDIX 4 - SUMMARY OF JT&S EFFLUENT SCHEMES IN THE MIDDLE EAST (1984)

Discussion on Papers 6 and 7

PROFESSOR M. B. PESCOD, Introduction to Paper 6
I would like to draw attention to the increasing importance of
food production in arid and semi-arid areas and the role urban
effluent reuse can play in agriculture. However, few urban
areas in developing countries are sewered and this is a prerequisite of wastewater reuse. Another essential factor in
planning reuse schemes is the co-operation and involvement of
the downstream user, in this case the agricultural sector.

Reviewing the health risks associated with effluent reuse I
suggest that most problems have been caused by the irrigation
with untreated sewage of vegetables eaten raw. Little
information is available on the health impacts of reusing
treated effluent in irrigation but my impression is that
health hazards are exaggerated. It is important in assessing
risk to consider the fate of pathogens, parasites, heavy
metals and other substances of concern during wastewater
treatment, in the soil and on the crop.

Contamination of groundwater is likely to be a concern in
many areas but I emphasize that the treatment capacity of the
soil should not be overlooked. Fig. 1, from Lance and Gerba
(ref. 1), shows the effectiveness of soil in removing viruses,
particularly under unsaturated flow conditions. Heavy metals
attenuation in sewage-irrigated soil is illustrated in Fig. 2,
from Evans et al. (ref. 2), which indicates an accumulation of
heavy metals in the surface layer. The long-term effects of
irrigating with sewage are clearly shown in Fig. 3, from Evans
et al. (ref. 2), and these environmental impacts must be
considered during the planning stages of effluent reuse
projects - not only heavy metal effects, which could have an
impact on health, but also the long-term effect of effluent
salinity on the permeability of the soil profile.

Water delivery, including effluent transport, storage and
application by irrigation systems, is of importance in
promoting the success of reuse in irrigation and this should
be an integral part of the planning process. Effluent
transport is costly and the siting of wastewater treatment
plants is a critical factor in the overall economics of reuse.
Storage requirements are dictated by the growing season but in
hot climates, where several croppings are possible in a year,

Reuse of sewage effluent. Thomas Telford Ltd, London, 1984

129

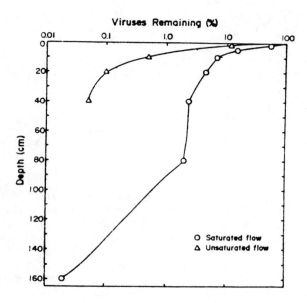

Fig. 1. Virus adsorption by a soil column during saturated and unsaturated flow. Saturated flow points are averages for three infiltration rates, and unsaturated flow points are averages for two infiltration rates

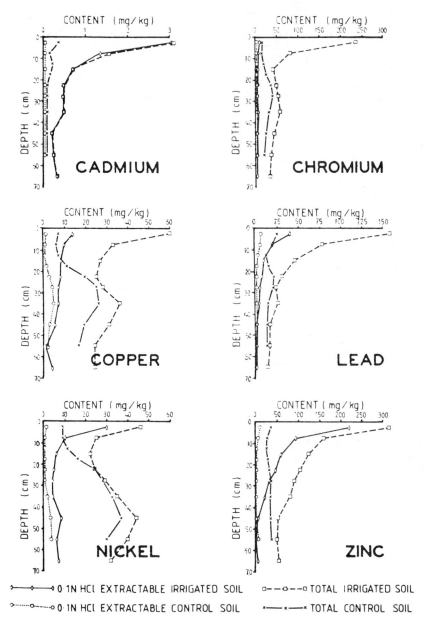

Fig. 2. The total and 0.1N HCl extractable heavy metal content of the soil profile

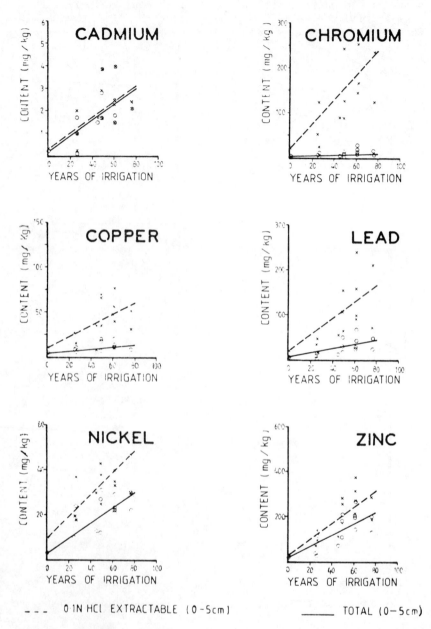

_ _ _ 0·1N HCl EXTRACTABLE (0-5cm)   _____ TOTAL (0-5cm)

Fig. 3. Apparent trends between metal content of soil and years of irrigation

careful choice of crops will allow a minimal investment in storage. In water short areas, effluent conservation must be promoted and the selection of irrigation system should be based on making maximum use of the limited resource. Charging for the treated effluent would be a convenient way of encouraging resource conservation in the downstream reuse.

Wastewater treatment for reuse in irrigation has generally been selected to achieve high effluent standards and typically involved conventional secondary sewage treatment plus tertiary treatment. Each project is unique and site-specific solutions should take account not only of the raw sewage characteristics but also the characteristics of the soil and the selection of crops. Stabilization ponds are known to be more effective in removing pathogens and parasites than conventional secondary treatment processes and are often more appropriate for adoption in developing countries. However, they are not always the optimal solution and it is essential to keep all options open in planning schemes.

There must be a more systematic planning of treated wastewater reuse in agriculture with the output objective allowing for social, economic and environmental factors. Project evaluation should take advantage of modelling techniques, such as multiple criteria analysis, in the attempt to arrive at the most appropriate integrated design. Further research, particularly on the epidemiology of effluent reuse in irrigation, is necessary and detailed monitoring of some of the projects now operating in the Arabian Gulf area should be undertaken.

REFERENCES
1. LANCE, J.C. and CERBA, C.P. Virus movement in soil during saturated and unsaturated flow. Applied and Environmental Microbiology, vol. 4, no. 2, 335-337.
2. EVANS, K.J., MITCHELL, I.G. and SALAU, B. Heavy metal accumulation in soils irrigated by sewage and effects in the plant-animal system. Progress in Water Technology, 1979, vol. 11, no. 4/5, 339-352.

DR J. E. BUTLER, Portsmouth Polytechnic, UK
Professor Pescod has said that conventional public health practice will not provide definite solutions in the field of reuse for agriculture and that we must be sufficiently open and flexible to respond to new, fresh approaches. This reflects a correct and refreshing attitude to the topic.

Professor Pescod and Mr Alka have rightly placed emphasis on the problems of pathogen survival and the possibility of soil contamination. I would be grateful if they would confirm that upgrading effluents by tertiary treatment may not solve all the problems because the use of rapid gravity sand filters and chlorination may still not prevent soil structure breakdown or contamination, and disinfection by lime dosing may create high pH values that few species of plant are able to tolerate.

Mr Cowan and Mr Johnson also highlight the importance of

133

assessing the public health risk and have accurately
summarized the problems to be surmounted as the prevention of:
consumption of contaminated material, insect bites, breathing
contaminated air, and body contact with contaminated material.

I seek the views of all authors on the merit of placing a
greater emphasis on the selection of crops that will undergo a
manufacturing process and so remove the need for tertiary
treatment. Examples include:

(a) grasses, reeds, etc. which may be used for energy
    production;
(b) sugar beet (some Middle Eastern countries are capable of
    producing their whole sugar need but have not had great
    success because of soil contamination problems);
(c) plants that can be used to produce pharmaceutical
    products, e.g. camomile, sunflower and castor to produce
    oils, and datura to produce heart drugs.

Professor Pescod said there was a need for new initiatives
and I would appreciate the opinions of all authors on the
approach being followed at Portsmouth Polytechnic. We have
taken the nutrient film technique and given it an 'appropriate
technology' transformation and created a system of irrigation
called gravel bed hydroponics (GBH). For several years it has
been possible to grow high quality fruiting crops such as
peppers, aubergines, melons, strawberries and tomatoes using
final effluents.

The nature of the GBH system removes three of those risks
listed in Paper 7. Also the system requires no soil; it
represents the most efficient use of water as there is no loss
in the ground and it is likely to suppress the helminth
problem.

Currently a co-operative scheme between Portsmouth
Polytechnic and the International University of Florida is
expected to exploit further the benefits of the GBH scheme.
(Funding from both UK and US sources of 450 000 US dollars is
expected.)

The papers identify a 'sea change' in government attitudes
over the last ten or twenty years to the reuse of sewage. The
impressive list of schemes mentioned bear witness to the
extent to which the authors of Papers 6 and 7 have played
major roles in creating this transformation of attitude.

Finally, I would welcome comments on a point made by
Professor Diamant that crop production using sewage effluents
should not be undertaken in an informal manner but that it
needs careful regulation and supervision. The Portsmouth
experience has been that correct operational procedures are
more important than initial design concepts.

DR A. ARAR, FAO, Rome, Italy
Papers 6 and 7 dealt with the potential use of sewage effluent
for irrigation in arid and semi-arid areas and quoted
examples, nearly all of them from the Near East region. In

this regard I would like to point out that the provision of
irrigation water is one of the most important factors for
increasing agricultural production. This is a very important
endeavour when it is remembered that more than 50% of the food
need of this region is being imported and that this percentage
will increase if the present trends of food production and the
demand for food continue.

Many countries in the region have utilized nearly all of
their conventional water resources (surface and underground)
and the future increase in irrigated agriculture has to depend
on other sources of water, such as desalination and treated
sewage water. However, the cost of desalinated water has
remained high during the last ten years owing to the rise in
cost of energy and it is more than 1 US dollar per m^3 which is
too high for irrigation purposes. On the other hand the reuse
of partially treated sewage water offers a good possibility.
It is encouraging to notice that many countries in the region
have plans in this respect and that some of these plans have
been implemented. Kuwait has been irrigating with sewage
effluent (secondary stage treatment) since 1977 and an area of
about 800 ha is being planted with forage crops, dry onions,
garlic and potatoes. Libya has a project of 750 ha irrigated
with tertiary treated sewage effluent and produces green
vegetables, e.g. lettuce, parsley, alfalfa and wheat. This
project has been going on since 1973 and it appears that they
have had no serious health problems, in spite of the fact that
sprinkler irrigation was being used on salad crops.

However, the above projects are based on tertiary treatment
which is costly and could not be adopted by non-oil-producing
countries for use in irrigation unless partially treated
sewage effluent with reasonable cost could be adopted. This
could be realized if the right crops are selected, e.g. forage
crops, cotton, sunflower, grain crops.

When discussing the effect of salinity on crop yield Paper 6
(Fig. 1) talks of the electrical conductivity of soil but in
paragraph 25 it quotes the salinity of sewage water for
irrigation, which does not distinguish between the salinity of
irrigation water and soil salinity. The criteria for salinity
in irrigation water should be based on several factors, namely
type of soil, type of crop, climatic conditions, washing
requirements and water management.

In semi-arid and arid conditions the soils are either sandy
or loamy with a high calcium carbonate content. However, in
Qatar they have very compact heavy and calcareous soils.
Therefore, the problem of sodicity is not expected to arise in
most conditions. Consequently, I am not convinced about the
wisdom of using lime, which is firstly costly, and secondly it
increases the salinity of irrigation water, to avoid a
sodicity problem. Hence the conclusions arrived at in Paper
6, in the case study of the UAE, should be reconsidered and
should be supported by research findings in the field and not
on assumptions. Paper 7 has reviewed in a comprehensive way
the plans for the treatment and use of sewage water for

irrigation in most of the Gulf states. However, the standards quoted for the salinity hazard are not up to date. No reference is made to the FAO irrigation and drainage paper 29 entitled 'Quality of water for irrigation' which discusses the chemical aspect of irrigation water. The soil, crop, climate and management interaction are also discussed for deciding on criteria for the chemical content of water for irrigation purposes. (Paper 6 made no reference to this FAO paper either.)

The heavy metal hazard in sewage water needs to be experimented upon to determine suitable standards for use in irrigation.

Industrial waste that contains harmful substances, such as heavy metals and complex organic compounds, which present serious health hazards should never be allowed to mix with the sewage effluent, in order to put an end to this rather serious hazard.

Both Papers 6 and 7 did not discuss the possibility of recharging untreated or partially treated sewage water to the groundwater to be pumped again for irrigation purposes. This aspect could prove to be one of the most suitable methods in the management of water resources and can be supplemented by cheap water from sewage effluent.

MR R. A. ANGIER, Howard Humphreys & Partners, Leatherhead, UK
Planned agriculture is essential. However, operation of treatment works is generally under one body and agricultural development using effluent reuse under another. What is the authors' experience of a combined management body?

Heavy metals in the surface of the soils are deleterious to crops. Does leaching help to reduce the problem?

For the feasibility of agricultural schemes using sewage effluent, apart from the benefits of crops, what is the value of effluent from the works?

Paper 7 mentions that drips and sprinklers are being used in Kuwait. Are there any problems with clogging and is special filtration needed?

In Abu Dhabi how were people enticed away from green verges and what were the health risks before they were enticed?

MR K. V. ELLIS, Department of Civil Engineering, University of Loughborough, UK
With regard to Paper 6 lime sterilization of effluents is too costly and too technically demanding for use in many developing countries. How effective is lime sterilization?

Chlorination of effluents is too costly, too technically demanding and too unreliable in its supply to be useful in the developing world. How effective is chlorination and what is the dosage rate, the availability of free chlorine and its effect on resistant bacteria, cysts, eggs and viruses?

In Paper 7 what evidence is there that sunlight is a major medium for sterilization in ponds (section 8)?

What evidence is there for effective removal of protozoan

cysts and worm eggs by tertiary treatment - particularly by
rapid sand filtration (sections 10 and 11)?

What is the source of information for Table 3?

I agree that the possibility of cross-contamination of
products on the way to the market (section 16) is of
importance, particularly if different qualities of water are
employed for irrigation of those products which must be cooked
before eating and those (salads) which are not cooked.

Who constituted the 'consensus' that 10/10 effluent is
always suitable for reuse (section 19)?

Are 10/10 effluents suitable for chlorination as they would
still contain a very appreciable chlorine demand and very
large doses would be required to produce free chlorine?
Without the presence of free chlorine how effective is the
chlorination of effluents?

MR J. LOUWE KOOIJMANS, Haskoning, The Netherlands
With the reuse of wastewater effluent for irrigation, two main
applications can be distinguished: applications for developed
countries and applications for developing countries. The
papers have shown that there is always money in the developed
world, including the rich Middle East, to apply advanced
technologies to the treatment of wastewater and reuse for
agriculture and other purposes.

However, for the developing countries the situation is
entirely different. A lot of people there have no access to
even a safe and reliable water supply and it is a first
priority to achieve this goal. Construction of sewerage and
sewage treatment systems, a prerequisite for reuse, has a
lower priority owing to a lack of funds and skilled staff for
operation and maintenance.

So, reuse of wastewater or wastewater effluent for
agriculture on a large scale in rural areas may not be
considered and it has to be restricted to those urban centres
where sewerage systems have been constructed and irrigable
land is available nearby.

It is clear from Paper 15 that reuse of raw sewage
introduces significant health risks for man and animals, and
this should be considered only under strict conditions and
only for some selected non-edible crops. Since advanced
treatment of wastewater will not be economically and
technically feasible (and is also unnecessary), we have to
look for low cost appropriate and reliable treatment
technologies, that can be applied in easy to operate and
maintain systems.

In their effort to contribute to the research for the
development of such technologies, the Dutch government
financed a pilot plant project in Cali (Colombia) in order to
investigate the application of the UASB (upflow anaerobic
sludge blanket) process for treatment of domestic wastewater
in developing countries under subtropical conditions.

The main advantages of this process are significant: low
investment costs (Fig. 4), low land requirement (Fig. 5), low

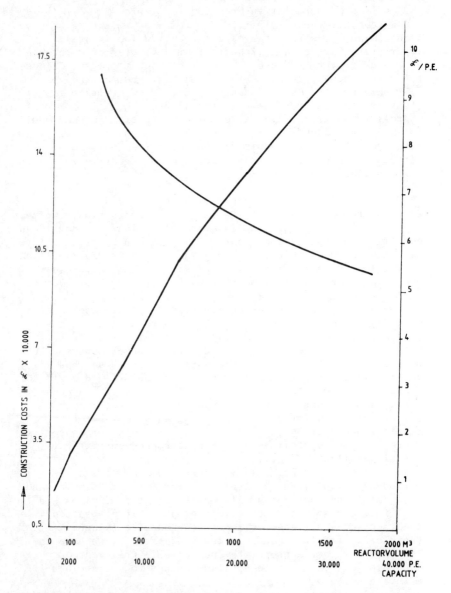

Fig. 4. Construction costs of UASB reactors under conditions in Columbia

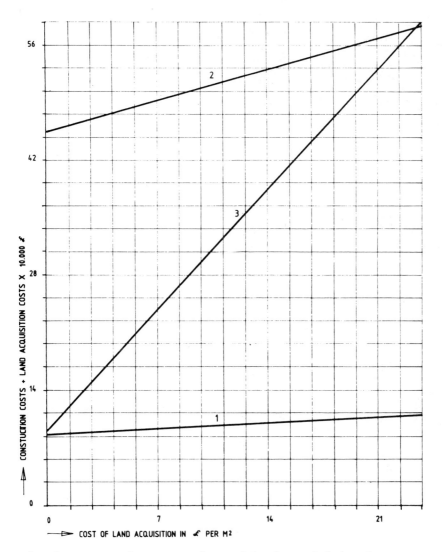

Fig. 5. Costs of construction and land acquisition for treat-ment plants with 16.000 PE capacity (1 = UASB reactor ($1020 \, m^2$); 2 = Carrousel ($5450 \, m^2$); 3 = Facultative lagoon ($21,000 \, m^2$)

LEGEND

1 SEWAGE COLLECTOR
2 SUBMERGIBLE INFLUENT PUMP
3 GRIT TRAP
4 FLOW CONTROL BOX
5 UASB REACTOR
6 GASMETER
7 SAMPLING PUMPS
7a REFRIGERATOR WITH SAMPLES
8 FEEDING PUMP FOR POST TREATMENT
   SYSTEMS
9 SEDIMENTATION TANK
10 TUBE PUMPS
11 TRICKLING FILTER
12 INTERMITTANT SLOW SAND FILTER
13 ANAEROBIC FILTER
14 MATURATION POND
15 DUCKWEED LAGOON
16 TO SEWAGE COLLECTOR
17 CHOICE BETWEEN SETTLED OR NOT
   SETTLED EFFLUENT

═══════ MAIN FLOW
─────── SMALL FLOW

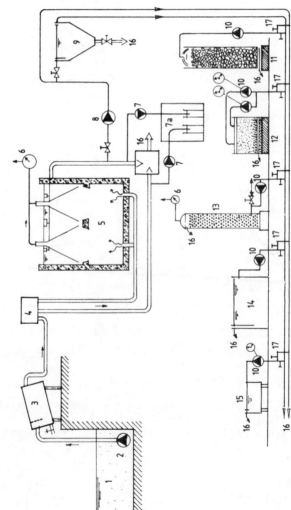

Fig. 6. Global layout of UASB pilot plant CALI with post-treatment systems

energy costs (just enough for transport to the biogas plant)
and low excess production.

In the upflow reactor (Fig. 6) a sludge blanket is formed,
in which anaerobic bacteria convert the biodegradable organic
matter via hydrolysis, acidogenesis and methanogenesis into
CO_2, H_2O, CH_4 and new biomass. Liquid, solids and biogas are
separated in the upper part of the reactor.

The research results were satisfactory, with 75-85% removal
of COD and 75-93% of BOD at a hydraulic retention time of 8
hours.

With respect to reuse for agriculture it is advantageous
that the nutrients nitrogen and phosphorus are not being
removed, but a disadvantage is that reduction of bacteria,
viruses, helminths etc. is low. For this reason and in
general to investigate reuse potentials, several post-
treatment tests have been carried out with, among others, a
trickling filter, an anaerobic postfilter, intermittent slow
sand filtration, lagooning and a duckweed lagoon. Although
the investigations are on going it may be concluded
provisionally that a UASB reactor combined with a maturation
pond (also required for storage) will give a reliable effluent
for reuse for irrigation.

MR J. HENNESSY, Sir Alexander Gibb & Partners, Reading, UK
Irrigated agriculture is a rural activity whereas treated
wastewater is an urban phenomenon in the Third World. What is
the perspective for areas served by treated wastewater? Why
do sewage treatment authorities not develop downstream
irrigated areas, under unified management, to grow commercial
crops? Would this not be a suitable development scenario to
show their longer term benefits to the local community?

MR K. FARRER, Watson Hawksley, High Wycombe, UK
In arid areas of the world, such as the Middle East,
irrigation demands can vary from 1.8 times the annual average
for the hottest parts of the summer down to 0.2 times the
annual average during the winter. The comparable variation in
wastewater flows during the same period could be 1.2 times
during the summer and 0.8 times in the winter, although this
varies with the location and nature of the wastewater being
treated. As a consequence, if maximum use is to be made of
the effluent for irrigation of permanent agriculture, such as
urban beautification, without long-term storage there is
likely to be a considerable shortfall in the amount of
effluent available to satisfy demands during the summer and a
surplus of effluent during the winter. The most common
methods for overcoming this problem are to suppliment effluent
supplies with potable water make-up during the summer and to
waste excess effluent flows during the winter or,
alternatively, to limit the amount of irrigation that can be
handled to the quantity of effluent available to meet summer
demands.

To illustrate the benefits that can be gained by storing

surplus winter flows for use during the summer and the magnitude of storage needed, I quote an example drawn from Watson Hawksley's involvement in the Phase II design of water and waste systems for the industrial city of Jubail on the east coast of Saudi Arabia. Jubail is a new city covering an area of approximately 450 km^2 with a projected resident population of 250 000 and planners have proposed a developing programme of urban beautification that will require ultimately some 90 Mm^3/yr of irrigation. It has two wastewater collection and conveyance systems; a sanitary system for the non-industrial areas (approximately 75% of the land area) with treatment at a sanitary wastewater treatment plant (SWTP) and an industrial system for the industrial areas (25% of the land area) with treatment at an industrial wastewater treatment plant (IWTP). On the face of it, the combined ultimate design flows of these two plants could provide sufficient effluent to satisfy most of the non-potable irrigation demands of the city. However, if maximum use is to be made of the effluent generated, long-term storage of some 17 Mm^3 would be needed to keep the quantity of potable water make-up to a minimum (approximately 5 Mm^3/yr).

The performance of the IWTP treatment processes has yet to be proved against actual wastes discharged by the petroleum-based industries at present under construction. If the effluent produced is to be used for irrigation it will consistently need to meet the stringent standards imposed for reuse before its full potential can be realized. Without IWTP effluent there will be a considerable shortfall in the amount of effluent available for irrigation, and the amount of surplus generated by SWTP during the winter for storage will be small when compared with the very high amount of potable water make-up needed during the summer (some 35 Mm^3 of make-up compared with 1 Mm^3 of storage).

Therefore, before deciding on the extent of beautification using reused effluent, it is vital that wastewater treatment be shown to meet consistently the prescribed standards for irrigation. Furthermore, it is essential for maximum use schemes that sufficient capacity be provided in the potable water system to be able to supplement or even to take over the irrigation on a reduced scale should the need arise.

In addition to treatment reliability, the economics of storage as compared to potable water make-up need to be considered fully before embarking on the development of a long-term storage policy. This is a particular problem in the Middle East where much of the supply comes from desalination plants and frequently the cost to the consumer is heavily subsidized. Other criteria which need to be established are actual variations in irrigation demand and wastewater flows during the year and, if possible, a realistic assessment of system losses for transmission and distribution. A very important factor for storage is whether and how to protect the very large volumes of stored water from evaporation which could be as high as 4000 mm/yr in some areas. All these factors have a considerable impact on the capacity of storage needed and hence the economic viability of proceeding with the storage policy.

With this in mind I would like to ask Mr Cowans and Mr Johnson whether any of the beautification schemes mentioned in their paper require potable make-up water during the summer months. If so, how much, and did they ever consider long-term effluent storage? Also, what use, if any, is made of surplus effluent generated during the winter?

DR J. D. SWANWICK, Sir M. MacDonald & Partners, Cambridge, UK
In response to Dr Butler's request for comments on his proposals for a new form of reuse of treated effluent in hydroponic culture, I refer him to work published in 1977 and 1978 in which lettuce and cucumbers had been successfully grown in experimental trials:
BERRY, W. L., WALLACE, A. and LUNT, O. R. Recycling municipal wastewater for hydroponic culture. Hort. Science, 1977, vol. 12, no. 3, 186.
WALLACE, A., PATEL, P. M., BERRY, W. L. and LUNT, O. R. Reclaimed sewage water: a hydroponic growth medium for plants. Resource Recovery and Conservation, 1978, vol. 3, 191.

PROFESSOR PESCOD and MR U. ULKA, Paper 6
In response to Dr Butler's contribution we agree that rapid gravity filtration and chlorination would not necessarily provide adequate preparation as tertiary treatment for municipal sewage from the point of view of the soil structure. However, recent tests have suggested that lime treatment need not necessarily produce an effluent with a high enough pH to affect plants. Storage would allow absorption of atmospheric CO_2 and can be expected to produce a gradual decrease in pH to a tolerable level. There is merit in producing crops subjected to processing after harvest but the downstream industry for high-technology products is not always available in developing countries. Whilst the approach being taken at Portsmouth Polytechnic in the development of the nutrient film technique is to be commended, it should be kept in mind that what is feasible in this country now will take some time to be manageable in most developing countries. The strict control and monitoring of any form of treated effluent reuse scheme cannot be overemphasized no matter how much attention to detail has been incorporated in the design process.

We thank Dr Arar for his comments and agree that further research on lime treatment is necessary before this approach can be accepted in any particular location, but the point in introducing the case study was to indicate that alternative forms of treatment to gravity sand filtration and chlorination could be effective and appropriate if the soil and crop requirements were considered. It is acknowledged that lime treatment might be costly in many countries but we recognize that economic comparison of alternative treatments should be an essential component of any feasibility study.

143

Mr Ellis extended this argument to include chlorination and questioned the reliability of this practice. We would prefer to rely on other forms of treatment, such as stabilization ponds, to remove pathogens and parasites but accept that chlorination will be applied in many countries.

Mr Angier raised the important question of institutional arrangements for effluent reuse in irrigation. We feel that either a single authority should handle the treatment and reuse of sewage effluent or very close collaboration between the sewage treatment agency and the agricultural user is essential. Lack of communication between agencies is too great a risk to public health. The answer to the question on the vertical mobilization of heavy metals in the soil profile will depend on the metal content of the applied effluent. If the effluent always contains the same levels of metals there will be a continual build-up of concentration in the surface layers of the soil but if the effluent is frequently devoid of metals there is a chance that the peak concentration of heavy metals will move down the soil profile. Finally, Mr Angier asked about the value of effluent and we can only suggest that this will vary from place to place, depending on the availability of alternative sources. However, it is felt that charging for effluent is a sensible approach which will result in most conservative use of the water and maximum agricultural productivity from the effluent available.

The authors thank Mr Kooijmans for his contribution and agree that low cost approaches to effluent reuse are essential for application in developing countries. Governments should, however, be made aware of the potential for urban effluent reuse wherever water supplies for irrigation are scarce and food production is important. Under these conditions, investment in wastewater collection and treatment might be economically justified.

In response to Mr Hennessy, we feel that planning for effluent reuse in developing countries requires the consideration of two factors:

(a) urban centres are usually surrounded by large areas of land which could be put into agricultural production or are already being farmed;
(b) there are, typically, large numbers of urban fringe slum and squatter dwellers, largely unemployed, who would benefit from the introduction of agricultural reuse of effluent which would provide employment within reach of the city.

Therefore, the question of the logistics of transporting effluent to rural centres for reuse does not normally arise. In many countries some form of direct or indirect reuse of effluent or raw sewage occurs in cities where sewerage systems have been installed. The issue then becomes one of ensuring that the safest and optimal methods of reuse are adopted to prevent risks to public health and/or environmental damage.

144

The management and organization of wastewater reuse must be tied to the available infrastructure in an individual country. However, it must be emphasized that close co-operation between the wastewater management agency and a number of government departments (particularly health, water and agriculture) is essential. Where individual farmer participation is planned, the wastewater treatment authority must be held responsible for the absolute quality of the effluent and such farmers must be strictly instructed as to the type of crops they must grow and how they should irrigate them. The assumption of responsibility for farming by the wastewater treatment authority is not necessarily feasible in all countries and it will generally be preferable to involve the agricultural sector as downstream users of a reliable effluent product.

Please refer to the addendum on p.330 for the replies by Mr Cowan and Mr Johnson

8 Constructive uses of sewage with regard to fisheries

R. J. HUGGINS, BSc, PhD, MIBiol, Wessex Water Authority, Poole, UK

SYNOPSIS   Sewage contains potentially valuable components which are difficult to extract. It may however be fed to fish either directly or through food chains. These processes allow some realisation of sewage as a resource because fish are valuable as animal protein and in some cases, such as weed control, they are operationally useful.

INTRODUCTION

1.   Sewage may be regarded as a potential resource and would soon become a real resource if the valuable components could be economically recovered. (ref. 1). In this context, Collinge and Bruce (ref. 2) assigned market values to some sewage sludge constituents and showed for example that the notional market values in one tonne of sludge dry solids (co-settled) were £64 for crude protein, £30 for fat, £10 for vitamin B12 and £3 for mineral oil (1979 prices). Unfortunately, extraction processes are often difficult and costly and may leave residues which require further disposal.

2.   Fish and other aquatic species are an acceptable and commercially valuable source of animal protein but modern intensive fish culture relies upon food derived from unpalatable protein sources such as fish meal. Such raw protein is becoming increasingly more expensive. The extraction of protein from sewage and its incorporation into fish foods therefore represents an attractive commercial proposition which is actively being investigated.

3.   The treatment of sewage frequently involves the use of lagoons and ponds where conditions sometimes allow fish to survive and grow. The production of fish in sewage lagoons is of interest both as an improved sewage treatment process (ref. 3) and for the production of useful fish species. Fish such as grass carp (Ctenopharyngodon idella) are being successfully used as an operation tool to economically control weed growth in watercourses and their production in sewage lagoons has been examined by fishery scientists of the Wessex Water Authority.

Reuse of sewage effluent. Thomas Telford Ltd, London, 1984

147

4.   Collinge and Bruce (ref. 2) noted that there may be
      scope for the utilisation of valuable components of
      sewage through food chains and Stirn (ref. 4) has
      proposed the use of sewage to increase marine
      productivity. The concept of increased production of
      edible marine species as an ecological consequence of
      marine treatment of sewage through long sea outfalls is
      currently receiving attention by marine scientists of
      the Wessex Water Authority. The following paper
      describes in some detail the investigations and
      concepts outlined above.

PROTEIN FROM SLUDGE - A FUTURE FISH FOOD?

1.   Edwards and Densem (ref. 5) noted that the
      supplementation of fish diets with protein extracted
      from sewage would offer a potential outlet for such
      protein if methods of extraction could be found. In
      recent years the Wessex Water Authority has been
      assisting a commercial group with the development of a
      novel method of protein extraction.

2.   The extraction method involves processing activated
      sludge through several physico-chemical stages. The
      product resulting from these processes typically
      contains 50% protein. In some respects the product
      resembles organic fertilizers such as dried blood and
      as such is currently being marketed in a fertilizer
      form. However, some preliminary experimental work on
      the incorporation of the product into fish diets has
      been carried out.

3.   In a small experiment, scientists of the Wessex Water
      Authority tested the palatability of the extracted
      protein to carp by replacing fish meal in fish food
      pellets with the sludge protein. Forty small mirror
      carp (Cyprinus carpio) were fed pellets in which
      approximately 50% of the fish meal component had been
      replaced by sludge protein. Carp which were fed food
      containing sludge protein readily accepted the diet and
      grew at rates equivalent to fish fed on commercial fish
      food based on fish meal. After 11 weeks of feeding on
      the food partly formulated with extracted sludge
      protein, the fish had grown satisfactorily but whole-
      body chemical analyses of the fish indicated some
      elevated levels of heavy metals.

4.   In this context, Tacon and Ferns (ref. 6) also reported
      significantly elevated levels of heavy metals in trout
      which had been fed diets containing activated-sludge
      protein. They believed, however, that heavy metals in
      sludge were not readily available as the food passed
      through the fish gut. Singh and Ferns (ref. 7) further
      investigated the accumulation of heavy metals in trout
      fed on activated sludge and concluded that the process
      of accumulation depended upon the metal involved and

that not all metals were accumulated. Recent work on the protein extraction process has significantly reduced the heavy metal content of the final product and with careful quality control it is possible that heavy metal contamination will not present a problem in the future.

SEWAGE LAGOONS AS FISH FARMS

1. The production of fish in ponds and lagoons which have been enriched with sewage has been a traditional form of aquaculture practised in the East for hundreds of years. The polyculture system widely used in China involves the enrichment of ponds with sewage nutrients and the subsequent rearing of several species of fish in the ponds. The process utilises the various levels of increased natural productivity such that filter feeding fish eat planktonic algae, weed-eating fish eat macrophytic growth, detritivorous fish feed in the mud and benthic feeders eat snails and shrimps.

2. In Eastern Europe, fish culture in sewage lagoons was common in the first half of the 20th century and, as an example, the carp rearing facility at the Munich sewage works is well known and widely reported. It is clear, however, that increasing industrialisation has resulted in sewage which contains greater amounts of toxic substances, many of which are persistent and accumulated in biota. Under these circumstances it is only at a few sewage works which have special facilities that fish culture for human consumption has continued (ref. 8).

3. Even with the problems of contamination, it remains clear that fish in sewage lagoons in temperate climates may still achieve very high growth rates which are comparable to those obtained in warmer climates where intensive culture techniques are applied (ref. 9). During studies of fish populations in sewage lagoons at Rye Meads sewage works (Hertfordshire, U.K.), Noble (ref. 10) noted that although contamination presented problems when fish for food were grown in the lagoons, coarse fish used for sport fishing could also be successfully grown in the lagoons. Contamination then ceased to be a problem because coarse fish comprise several species which are used for sport but are not eaten, being returned to the water after capture.

4. The concept of using sewage lagoons for the rearing of useful fish which are not immediately destined for human consumption warrants further examination and in this context, Allen and Gearheart (ref. 11) have reported on the use of wastewater lagoons for the rapid low-cost rearing of young migratory salmon as part of an ocean-ranching aquaculture system. The young coho salmon are reared in wastewater lagoons before being

released to spend a minimum of 4-6 months free-ranging in the ocean. They then return to their natal stream as large adult fish. Allen and Gearheart were confident that adult fish which had been initially reared through juvenile stages in wastewater lagoons would meet public health regulations.

5. Henderson (ref. 3) has reported on the use of fish to improve the quality of effluent in wastewater lagoons again using filter-feeding chinese carp which were commonly used in polyculture systems. The fish were stocked in the last four ponds at a six-pond wastewater treatment facility where significant reductions in BOD and suspended solids were achieved. The filter-feeding fish affected algal populations in the ponds and in some way stabilized fluctuating levels of dissolved oxygen.

6. The enhanced production in wastewater ponds of useful fish which are not destined for human consumption led to experiments by fisheries scientists of the Wessex Water Authority with grass carp in sewage lagoons. Grass carp (Ctenopharyngedon idella) are weed-eating fish which are frequently used in chinese polyculture systems. They are of considerable economic interest to the Wessex Water Authority because they may provide a cheap, alternative method of weed control in some watercourses which are currently cleared of weed by expensive and potentially damaging chemical and mechanical methods. The fish, which are not indigenous to the U.K., do not breed in U.K. rivers and must be artificially bred and reared to a minimum stocking size. It was therefore decided to experimentally rear grass carp in a hectare lagoon at the Avonmouth sewage treatment works near Bristol, U.K. and in November 1977, 1000 small (10cm) fish were introduced into the newly constructed lagoon which had been filled with sewage effluent.

7. Throughout the winter of 1977/78 the un-ionized ammonia levels in the lagoon were found to be very high and in March 1978, 365 fish were rescued from the lagoon after showing signs of distress. When measured, these fish showed that no growth had taken place in the six-month winter period and in fact some loss in weight and condition had occurred. In July 1978, 264 fish were returned to the lagoon and no further examinations took place until June 1980 when it was necessary to drain the lagoon. 84 fish were rescued and these, on average, had increased from the November 1977 weight of 20.2gm to 243.3gms. The surviving fish were found to be suffering from a disease which caused skin lesions. Even with the environmental problems of disease and chemical stress, the fish had increased their average weight by 10-fold in 31 months which included two

winter periods. In further studies, disease-free grass carp have been stocked into wastewater lagoons with lower ammonia levels and when introduced into the ponds in early spring the fish have grown well throughout the summer and thus attained sufficiently good condition to enable their subsequent survival through the winter period.

8. In conclusion it is interesting to note that Allen (ref. 12) regarded sewage fish-pond systems as highly productive but relatively unstable ecosystems lying half-way between controlled laboratory experiments and eutrophic natural environments. Recent results are encouraging in that by careful design and control, sewage fish-pond systems may yet be used to produce food fish, useful fish and to achieve improvements in effluent quality. The traditional chinese polyculture systems may yet further contribute to the modern biotechnology of aquaculture in wastewater.

MARINE TREATMENT - POLLUTION OR BIGGER CATCHES?

1. At some sites, toxic wastes have been released into the sea in sewage and dramatic and emotional descriptions of the resulting environmental damage has led the public, and in particular the inshore fishing community, to regard all marine disposal of sewage as unacceptable pollution. The scientific reality is, however, somewhat different. In fact, where domestic sewage, free of toxic industrial contamination, is released into the sea through properly designed and sited outfalls, the most common ecological response involves increased productivity and in many cases an increase in species.

2. Early experiments, where fertilizers were applied to sea lochs, showed remarkable increases in plankton and bottom fauna (refs. 13, 14). In this context it is interesting to note that James and Head (ref. 15) in a study of the discharge of the River Tyne into the sea concluded that the river, which comprised up to one third sewage effluent, increased nearshore marine planktonic production with few, if any, detrimental effects. Stirn (refs. 16, 4) developed the idea of a constructive use of sewage for increased marine productivity but sadly this prospect has received little further attention probably because the environmental revolution of the 1970's regarded the disposal of any waste into the sea as unacceptable.

3. It is clear from an examination of recent reports that where dispersion of sewage into the sea from a well designed and sited outfall is used as a method of wastewater treatment, increased marine productivity is frequently evident. In fact some marine scientists seem to be surprised that sewage in the sea can have

151

beneficial effects. Caspers (ref. 17) after investigating the ecological effects of dumping sewage sludge from Hamburg into the German Bight expected to find a sterile and depopulated seabed. His samples showed, however, a rich macrofauna to be present and Caspers was further surprised to observe that the more sludge dumped, the denser was the population of some shellfish.

4. After ten years of studying the effects of sewage discharges into southern California coastal waters, Bascom (ref. 18) concluded that fish and epibenthic invertebrates have increased in biomass, numbers of individuals and number of species. Detailed studies of a new long sea outfall for Edinburgh into the Firth of Forth have also shown the same ecological responses (ref. 19).

5. Studies of an existing outfall at Bridport, U.K. by marine scientists of the Wessex Water Authority again showed an increased marine productivity around the discharge with more species and more individuals of some species when the discharge site was compared to distant control sites. Such increases in species and productivity were manifest to scuba divers as an underwater "oasis" around the outfall and local fishermen know of the localised increase in marine productivity. Studies of their fishing habits has revealed that they take advantage of the ecological effects of the outfall by carefully deploying their crab traps near to the outfall.

6. It is suggested that a co-operative effort involving local fisheries interests, marine fishery scientists and the wastewater management authority should be considered when any new outfall is proposed. Through such a project, the fertilising effect of sewage in the sea could be utilised for the benefit of local fishermen. The concept assumes that increased marine productivity as a result of the wastewater discharge will occur but it recognises that without environmental modification, much of the productivity will be evident as non-edible, non-commercial species. By carefully designed fishery management techniques it is proposed that the increased productivity should be manipulated to produce commercially important species.

7. During studies of the newly constructed outfall for Weymouth and Portland U.K. it was clear that the presence on the seabed of the outfall structure was sufficient to attract some species even before wastewater discharge took place. It is also interesting to note that the large fish populations around wrecked ships on the sea bed are well known to sport fishermen. Work in California in the 1960's (ref. 20) showed that artificial reefs could

substantially increase fish populations and Shelbourne (ref. 21) suggested the augmentation of local fish stocks by the provision of protected seabed areas which had been enriched by fertilisation.

8. It is therefore possible to conceive of a project in which a wastewater discharge into the ocean would be combined with a marine fish farming development. Artificial reefs would be constructed around the discharge point and at various distances from it. Such reefs would be designed to increase survival and growth of young fish and crustaceans while shellfish such as mussels could be rapidly grown using traditional mussel farming techniques in the area of increased phytoplankton production. Such a concept involving the beneficial use of sewage to improve marine inshore fisheries would be dependant upon two factors. Firstly, there would need to be interest from marine fisheries scientists and commercial fisheries interests and secondly, strenuous efforts would be necessary to exclude peristent toxic substances from the discharged wastewater.

THE FUTURE

1. This paper is presented in an attempt to fire the imagination of those engineers who, because of public criticism, may regard the disposal and treatment of sewage as a difficult environmental problem with few, if any, options for constructive and beneficial development. Within existing scientific knowledge, information already exists which would allow for the constructive use of sewage particularly with regard to fish and fisheries. It is hoped that the defensive attitude which is commonly taken, especially when marine treatment of sewage is proposed, will disappear and that an optimistic view of sewage as a resource will emerge with particular regard to fish.

REFERENCES

1. COLDRICK, J. Sewage as a resource. New Scientist 30 Oct. 1975. 276-278

2. COLLINGE, V. K. and BRUCE, A. M. Sewage sludge disposal: a strategic review and assessment of research needs. Water Research Centre Technical Report TR166. 1981

3. HENDERSON, S. Utilisation of silver and bighead carp for water quality improvement. In Aquaculture Systems for Wastewater Treatment. E.P.A. Seminar Proceedings, Davis California. September 1979

4. STIRN, J. Possibilities for the constructive use of domestic sewage (with an example of the Lake of Tunis). In Marine Pollution and Sea Life (M. Ruivo, Ed.), Fishing News (Books) Ltd., 1972, London

5. EDWARDS, R. W. and DENSEM, J. W. Fish from sewage. In Apphed Biology Vol. V, 221-270 (T. H. Coaker, ed.) Academic Press 1980

6. TACON, A. G. J. and FERNS, P. N. The use of activated sludge from domestic sewage in trout diets. Nutr. Rep. Int. 13 (6) 549-562. 1976

7. SINGH, S. M. and FERNS, P. N. Accumulation of heavy metals in rainbow trout maintained on a diet containing activated sludge. J. Fish. Biol., 13 (2) 277-286. 1978

8. HUGGINS, T. C. and BACKMANN, R. W. Production of channel catfish in tertiary treatment ponds. Iowa State University, Project A-017-1A, 1969

9. WHITE, R. W. G. and WILLIAMS, W. P. Studies of the ecology of fish populations in the Rye Meads sewage effluent lagoons. J. Fish. Biol., 13 (4) 379-400. 1978

10. NOBLE, R. P. Growing fish in sewage. New Scientist, 31 July 1975. 259-261

11. ALLEN, G. H. and GEARHEART, R. A. Public health aspects of a wastewater-based California salmon ranching project. CSIR Symposium on Aquaculture in Wastewater, Pretoria, 24-26 November 1980

12. ALLEN, G. H. The constructive use of sewage with particular reference to fish culture. In Marine Pollution and Sea Life (M. Ruivo, ed.), Fishing News (Books) Ltd., 1972, London

13. MARSHALL, S. M. An experiment in marine fish cultivation. III. The plankton of a fertilised lock. Proc. roy. Soc. Edinb. 63 (B) 21-33, 1947

14. MARSHALL, S. M. An experiment in marine fish cultivation. IV. The bottom fauna and food of flatfishes in a fertilised sea-lock. Proc. roy. Soc. Edinb. 63 (B), 34-55, 1947

15. JAMES, A. and HEAD, P. C. The discharge of nutrients from estuaries and their effect on primary productivity. In Marine Pollution and Sea Life, (M. Ruivo, ed.), Fishing News (Books) Ltd., 1972, London

16. STIRN, J. The consequences of the increased sea bioproductivity caused by organic pollution and the possibilities for the protection. Revue int Oceangr. med., 10 123-129, 1968

17. CASPERS, H. Ecological effects of sewage sludge on benthic fauna off the German North Sea coast. Prog. Wat. Tech. 9 (4) 951-956, 1978

18. BASCOM, W. The effects of waste disposal on the coastal waters of southern California. Environ. Sci. Technol. 16 (4) 226A-236A, 1982

19. READ, P. A., ANDERSON, K. J., MATTHEWS, J. E., WATSON, P. G., HALLIDAY, M. C. and SHIELLS, G. M. Effects of pollution on the benthos of the Firth of Forth. Mar. Poll. Bull. 14 (1) 12-16, 1983.

20. CARLISLE, ·J. G., TURNER, C. H. and EBERT, E. E. Artificial habitat in the marine environment. The Resources Agency of California, Dept. Fish and Game. Fish Bulletin 124. 1964

21. SHELBOURNE, J. E. The artificial propagation of marine fish. Adv. Mar. Biol., 2 1-83. 1964

I wish to thank K.F. Roberts, C.B.E. (Chief Executive, Wessex Water Authority) for his kind permission to present this paper. The views expressed in this paper are not necessarily those of the Wessex Water Authority.

9 Use of sewage waste in warm water aquaculture

A. I. PAYNE, BSc, MSc, PhD, Coventry (Lanchester) Polytechnic, UK

SYNOPSIS. Ponds receiving sewage effluent can show rates
of fish production from 2-6t ha^{-1} yr or occasionally higher.
Algal production is also great and in high rate stabilization
ponds whole total production can attain 110t ha^{-1} from the
microflora. The presence of both fish and algae can also
considerably improve the quality of an effluent by reducing
BOD, COD, nitrogen and to a lesser degree, phosphorus. The
extent of this improvement depends upon the system adopted.
There is also considerable attenuation of faecal coliform
bacteria by several orders of magnitude within ponds. The
input of sewage into fish ponds must be controlled to maintain
environmental conditions such as oxygen and ammonia at levels
which the fish will not only survive but which will also not
inhibit growth. Bioaccumulation of toxic trace compounds and
the role of fish in transmitting human diseases must be taken
into consideration.

SOME GENERAL PRINCIPLES
Sewage and aquaculture production

1. The use of human waste in the culture of fish is not a
new idea. Many of the traditional methods developed in the
Far East, which currently produces more than half the world's
total of 6 million tonnes of farmed fish, uses night soil or
human waste in some form to promote fish productivity and
latrines may even be built directly over fish ponds. This,
of course, is just an extension of the general principle of
using animal manures in both agriculture and aquaculture and
certainly, as far as fish are concerned, the use of organic
manures to enhance production has become sufficiently developed, under some circumstances, to sustain fish yields in
excess of 7t ha^{-1} (ref. 1; ref. 2).

2. However, domestic animal manures are intended specifically as a support for fish culture as a primary objective,
whilst the introduction of fish farming into sewage processing
and disposal is as a beneficial adjunct which might convert
some of the waste into a useful product and also may improve
the efficiency of the sewage treatment.

3. There are two ways in which organic manures influence
fish production, both indirectly through mineralisation of
the organic material to provide inorganic nutrients for the

Reuse of sewage effluent. Thomas Telford Ltd, London, 1984

157

algae of the phytoplankton, which can then be used as a source
of food by appropriate fish species, and through direct consumption of the waste as food. The provision of treated domestic waste introduces an immediate supply of inorganic plant
nutrients, such as ammonia and phosphate, for algae since
primary treatment generally means that a certain amount of
decomposition and mineralisation has already occurred. The
laying out of water in sewage oxidation ponds provides further
opportunity for this process to continue, consequently blooms
of algae of considerable densities arise in these circumstances. Moreover, domestic sewage often contains a high phosphate concentration owing to the large amounts contained in
detergents. A simple chemical model of primary production(ref.3)
in approximately stoichiometric proportions is provided by:

$$122CO_2 + 16NH_4^+ + PO_4 + 58H_2O \xrightarrow{\text{Light}} C_{122}H_{179}O_{44}N_{16}P + 131O_2 + H^+$$
$$\text{algal biomass}$$

The carbon source can either be derived from the carbon
dioxide produced from the decomposition or from the carbonate
and bicarbonate reserves in the water. The algal biomass
itself can comprise 50-60% protein by dry weight which makes
it potentially a very valuable feed.

4. The rate of mineralisation is rather more rapid in
warmer climates and this, through the nutrient supply, has an
equivalent effect on algal production, so that to maintain an
algal concentration of 300mg l^{-1} in Israel required a pond
detention time of 7-8 days in the winter at a mean temperature
of 13.6°C but only 1.8 days in the summer when average temperatures were 25.4° (ref. 3). In the more uniformly warmer
regions of the tropics shorter detention time periods are,
therefore, to be expected. The large algal production from
sewage oxidation ponds or those fed by effluents can be
utilized either by directly stocking fish into these, if they
can tolerate the difficult environmental conditions that
occur, or the algae can be harvested by filtration, flocculation or centrifugation to be used as a protein rich additive
to compounded diets for fish or other domestic animals.

5. In addition to the algae, the fine particles of organic
material derived directly from the sewage and their associated
bacteria can be utilized as found by some fish either indirectly via the zooplankton or directly if the species used can
extract fine particulate matter. Furthermore, organic compounds in solution may also be used by both phytoplankton,
particularly blue-green algae, where it may account for up to
20% of total production (ref. 3) and bacteria as an additional
resource for production.

6. The types of fish which can best profit from sewage
waste are those which habitually feed upon small particles or
upon bottom sediments. Amongst these are some of the carps
such as the chinese silver carp, Hypophthalmichthys molitrix
and the European common carp, Cyprinus carpio or the tilapias
largely originating from Africa. These fishes have been used

extensively for warm water aquaculture although within the
tropics the temperature regime may approach the upper lethal
limits for the carps. Amongst the environmental limitations
imposed by the sewage itself the most significant is the oxygen
level, which may generally be low due to the high BOD of the
effluent or show wide fluctuations owing to the algal blooms.
Very low oxygen concentrations will kill fish but even moderately low values will inhibit growth and production. The same
is also true of ammonia which characteristically attains high
concentrations in sewage effluent.

7. The role of aquaculture in sewage treatment. Amongst the
aims of sewage treatment are the reduction in suspended solids
and the BOD associated with the organic component, to improve
oxygen relations in the water and to avoid an excessively acid
pH. The decomposition of the organic matter gives rise to
inorganic nitrogen containing compounds which, if released into
a natural water body, will promote excessive production in
natural communities to give the condition known as 'eutrophication', the symptoms of which cause a decline in the water
quality. Moreover, even oxidised forms of nitrogen in the form
of nitrites and nitrates can constitute a direct health hazard.
For example, excess nitrate above 22 mg l^{-1} may even be lethal,
particularly to infants, due to the formation of methaemoglobin
in the blood, which is a serious consideration if the effluent
is to be mixed with water to be used for drinking.

8. An incorporation of some form of aquaculture system at
some stage of the processing of the sewage or the effluent has
the potential to improve some or all of these features. The
inorganic nutrients containing phosphorus and nitrogen become
incorporated into algae when detained in ponds or tanks, as
outlined in the above equation, and promote the development of
considerable densities of algae. If these are left untouched
they can cause large variations in the oxygen levels in the
ponds such that the water becomes supersaturated in the evening
and almost anaerobic in the early morning. Blue-green algae
which can be common in such ponds, can cause surface scums
which interfere with re-aeration, whilst the die-off underneath
the algae scum can further deplete the oxygen. Extra-cellular
products from the algae may render the water undrinkable or
unsuitable for re-use, whilst the high algal densities can clog
up any subsequent downstream filter beds or screens. Removal
of the algae by plankton feeding fishes concentrates a proportion of the nitrogen and phosphorus which would otherwise
contribute to eutrophication, into the body tissues to be removed from the system when the fish are harvested. In addition,
the fish may also remove directly some of the fine particulate
organic material.

9. Harvesting the algae directly would also eliminate the
accumulated excess phosporus and nitrogen but although this is
technically feasible it tends to be relatively expensive.
Harvested algae can be used as a fertilizer on the land or as
a supplement to animal feeds, although it is perhaps not so
immediately utilizable as a fish crop. Removal of the algae

by either of these methods, however, significantly contributes to tertiary treatment of the water. Controlled production of algae is also of value in improving the oxygen relations of the pond providing oxygen output from photosynthesis in the day exceeds utilization by respiration by respiration by algae, bacteria and animals during the night. If photosynthesis is very rapid and requires carbon from the base reserves of the water, as it usually does, then the pH of the water will rise during the day, even to levels of pH9 or 10, thus offsetting the general acidifying effects or organic decomposition.

10. One further benefit of an aquaculture approach to sewage treatment is the appreciable disinfection effect the community production system found in fish ponds appear to have on coliform bacteria including the enteric on pathogenic types. This in itself is of considerable value, where the possibility exists of contact between man and the final effluent.

THE USE OF SEWAGE EFFLUENT TO PROMOTE AQUACULTURE PRODUCTION
Open waters

11. The release of sewage into water creates a considerable demand for oxygen due to bacterial respiration as the organic material is broken down. In the tropics and sub-tropics this demand for oxygen can be considerable as indicated by the high range for chemical oxygen demand(ref.4) which had been recorded of 800-1400 mg l^{-1} compared to countries such as the USA where 400-500 mg COD l^{-1} are more prevalent. The reason for this is partly due to the more incomplete treatment of sewage in tropical countries but also reflects the lower volumes of water passing through the treatment. The effects, however, can be quite striking, for example most of the sewage from Cairo, Egypt is released into the Bahr-el-Baquar Drain which carries it to one of the northern coastal lakes, Lake Manzala, some 140 Km distant. Even at a point close to its entrance to the lake the water is black with suspended material and close to being anoxic with only 0.2% saturation during the afternoon when small tilapias, which are generally tolerant of low oxygen, can be seen right at the surface respiring at the air/water interface (ref. 5). Where the water from the drain enters the south-eastern corner of the lake a dense algal bloom is formed over the whole area and estimates of fish yields, mainly plankton feeding tilapia, from this sector of some 441 Kg ha^{-1} are seven times greater than those from other parts of the lake (ref. 6). This demonstrates firstly the propensity of sewage to promote fish production and also that the worst effects of eutrophication, which arises when the partially mineralised sewage reached the oxidising conditions provided by the shallow unstratified lake, appear to have been confined to the south-eastern corner. This may reflect a relatively rapid completion of decomposition and nutrient recycling in this warm environment since an effluent inflow of this magnitude would probably have had rather more widespread effects in a temperate lake; there are also no phytoplankton feeding fishes in many temperate areas. Fryer (ref. 7) similarly noted that the

whole sewage output from the sizeable town of Entebbe, Uganda was discharged into Lake Victoria but the effects were very localised. Nevertheless, the effects from eutrophication can cause substantial changes to the biological community which in turn can impair ability to purify subsequent additions of sewage. So that although fish production can be encouraged, the uncontrolled discharge of effluent into natural waters does carry significantly high risks.

12. Reservoirs can be used in a similar way, for example a reservoir of 4 ha in Israel which received the sewage from a town of 5,000 people, produced a fish yield of 2,000 Kg ha^{-1} over a six-month period, which is 10 to 100 times the natural productivity of such waters. Alternative uses for the water which has been treated in this fashion does, however, become restricted.

Production from sewage-fed ponds

13. Fish can be considered for use with sewage either during the initial stages of treatment or through the use of effluent water in which is effectively tertiary treatment. The use of raw sewage generally creates too many environmental problems. The simplest form of treatment using sewage oxidation where the sewage/water mixture is left for an appropriate length of time for biological purification by the processes outlined in the opening section, lends itself particularly well to combination with aquaculture.

14. In a fairly typical example of a small-scale unit receiving a loading of 49.6 Kg BOD$_5$ ha^{-1} day^{-1} from a community of 2,300 individuals, in which the sewage passed through a bar screen and grinder, a clarifier, and anaerobic digester before finally flowing into a series of three oxidation ponds (ref. 9), it proved impossible to maintain mixtures of silver carp (Hypophthalmichthys molitrix), bighead carp (Aristichthys nobilis) and grass carp (Ctenopharygodon idella) in the first pond of the series owing to the poor water quality of the raw effluent. In the second pond in the series the fish could maintain themselves but production was poor at 454 Kg ha^{-1} and there were occasional fish kills. In the final pond, however, there was 84% survival and a combined yield of 2,729 Kg ha^{-1} of which 2,554 Kg ha^{-1} was the phytoplankton feeding silver carp. It was also notable in this case that the algae which predominated were small greens such as Chlorella, a common feature of oxidation ponds in both warm and temperate climates (ref. 10, ref. 11) whilst in control unstocked ponds run in parallel the principal types were blue-greens with their tendency to form scums and reduce oxygen through die-offs. The grazing of the fishes also appeared to stimulate algal growth. As a result of this procedure, not only was a substantial fish crop obtained but there were also significant improvement in water quality as outlined in the following section.

15. In a similar system Suffern et al (ref. 12) kept all-male hybrid tilapia (Oreochromis mossambicus x O. hornorum) although in this case the fish were kept in cages within a series of

ponds, the first of which received sewage which had only been
passed through a mechanical chopper and possessed a BOD_5 of
79 mg l^{-1}. The ponds were, however, fitted with bubble aera-
tors which managed to keep the oxygen content of the ponds above
5.5 mg l^{-1} (approximately 60% saturated). The fish grew well in
both ponds and taking into account the scale of the experiment
and probable limits of algal production a practicable unit fish
production for the system was estimated to be 50t ha^{-1} yr^{-1}.
The use of cages with both tilapias (Oreochromis mossambicus) and
silver carp in sewage effluent has also been successfully demonstrated in Southern Africa (ref. 13).

16. The integrated use of warm water and sewage effluent to
promote algal production to grow planktivorous fishes is also a
possibility in temperate areas (ref. 14). Controlled input of
60 Kg organic waste per hectare per day into primary treatment
ponds kept at between 15-30°C prompted the growth of 80 Kg ha^{-1}
day $^{-1}$ phytoplankton which led to an average production of 35 Hg
ha^{-1} day tilapia hybrids over a 180 day cycle leading to a final
yield of 6.3t ha^{-1} based only upon the organic waste input alone.

17. In India the use of sewage effluents in fish ponds has
become quite widespread and in West Bengal, for example, some
80,000 ha of ponds receive sewage waste (ref. 15). Fish production in these ponds, mainly from the Indian major carps, has
reached 2.3t ha^{-1} with some using silver carp, attaining 7.2t
ha^{-1} and with tilapia, 9.4t ha^{-1} (ref. 15). In Israel sewage
from a Kibbutz of 500 people fed into fish ponds and also
receiving supplementary feed produced 8.6t ha^{-1} over eight
months, whilst adjacent ponds receiving only chemical fertilizers and similar feeding yielded only 4.7t ha^{-1} (ref. 8).

18. China also has an extensive fish farming system which is
dependant upon organic manures including human waste. This,
however, is not now used directly and is normally allowed to
ferment for four weeks in closed chambers before use (ref. 16).

19. A potentially important intensive system has recently been
devised in Taiwan (ref. 17) which utilises octagonal tanks of
$100m^2$ area into which the fish, normally 'red tilapia' which are
hybrids of up to four species (ref. 18) are stocked at very high
densities. Part of the water is replaced twice daily with
enriched water and the pond is kept aerated and fine particles
kept in suspension, by paddle wheels. The process is rather
similar to activated sludge treatment and the yields are remarkably high 3-4 tons/pond/crop or 6-8 tons/year being obtained.
This has not actually been used in conjunction with sewage
effluent but the principles are the same.

20. Generally ponds receiving only inorganic fertilizers but
with supplementary feeding can attain yields of 1-2t ha^{-1} yr^{-1}
whilst intensive systems using mixed fish species with large
inputs of animal manure can produce up to 8-10t ha^{-1} yr^{-1} and
possibly higher with aeration (ref. 19). In these terms ponds
receiving sewage effluent can produce high yields, although
perhaps not the highest under most field conditions, owing to
the sub-optimal environmental conditions. However there is considerable scope for improvement by using sewage under controlled

conditions which depart as little as possible from the optimum
for the fish.

High rate stabilization ponds

21. Given the major role that algae can have in the processing of sewage (ref. 20) systems have been devised to maximise
the use of sewage by algae to improve the efficiency of the
process and also to obtain the highest algal yield (ref. 21,
ref. 22, ref. 23, ref. 3). Providing scum forming algae can
be avoided the algal production will provide some of the beneficial effects on water processing mentioned above, whilst the
production itself can be harnessed by using herbivorous fishes
to graze down the phytoplankton blooms or the algae themselves
can be harvested direct. The process is most effective in
warm climates with high incident solar radiation to promote
rapid photosynthesis.

22. The principal of the method is to feed sewage into a con-
voluted series of channels created from a series of baffles
within a tank and to continuously circulate this
through the channel circuit where, in strong sunlight, the rich
algal bloom will develop. The depth of water is maintained at
35-50 cm to allow light penetration through a significant proportion of the water column and the water is driven by paddle
wheels or some form of rotary aerator which also keeps the
algae in suspension. It should, however, allow a sedimentary
bacterial phase to develop on the channel floor to promote
decomposition of the organic material prior to release of
nutrients used by the algae. Effective detention time can
range from 0.5-1.5 days as has been used in Thailand (ref. 24)
or up to 7.8 days in Israel during the winter (ref. 3) since
the process is light and temperature dependant although the
sewage loading must also be taken into account.

23. Production in these ponds can be very high. In Israel
total production of all material amounted to an average of 39 g
day weight m^{-2} day^{-1} (ref. 3) of which 22 g m^{-2} was due to the
suspended algae. Against this last value, and bearing in mind
the greatest rate of algal production recorded, in June, was
37.2 g m^{-2} day^{-1}, can be the theoretical maximum for photo-
synthetic algal production. 32 g Carbon m^{-2} day^{-1} (ref. 24)
which can be approximately converted to 57 g dry wt. m^{-2} day^{-1}
thus demonstrating just how high production can be in these
ponds. Similar values for total production values from
Thailand have ranged from 15.7 to 39.3 g.m^{-2} day^{-1}, which are
equivalent to 57.3 to 112 tons ha^{-1} day^{-1} (ref. 10, ref. 25).

24. To utilize this tremendous productivity the algae can be
used to support fish production (ref. 26, ref. 27), it can be
harvested directly (ref. 3, ref. 28) or it can be used on the
land as a fertilizer (ref. 29). Using relatively weak domestic
sewage with BOD_5 of 45 mg l^{-1} Edwards et al (ref. 29) were able
to obtain a yield from the tilapia Oreochromis niloticus of
6t ha^{-1} yr^{-1} in ponds fed from a high rate stabilization pond
in Thailand. However, the efficiency of the high rate pond at
producing algae was greater than that of the fish at harvesting

and consequently many algae passed through the fish ponds and needed to be dealt with by an additional means. In this case the water from the fish ponds were put into maize plots which also benefitted to produce a substantial yield. From the values provided from this study it was estimated that to process the effluent from a town of 100,000 people in the tropics would require 8.94 ha of high rate stabilization pond, 4.8 ha of fish ponds and 49.2 ha of maize. The contribution by aquaculture would therefore be relatively small. The general economic prognosis for the whole system was not very favourable and led to a conclusion that biological harvesting of algae from a high rate pond is difficult to integrate and, therefore, either the algae should be harvested directly or the use of conventional oxidation ponds with fish was more practicable.

25. The algae can be separated from the water by flocculation with aluminium sulphate $(Al_2(SO_4)_3)$ or ferric chloride $(FeCl_3U_3)$ at dosages of 70-120 mg l^{-1} followed by flotation and centrifugation (ref. 3). In a dried form this can form a valuable basis for animal feeds and has also been used successfully with herbivorous fishes such as the grass carp (ref. 30). Even with the presence of 4% aluminium in the harvested algae from the high rate pond 30% dried algae in the diet of the common carp and the tilapia, Sarotherodon galileus, could be used to replace 85% of expensive fishmeal in pelleted artificial feeds (ref. 31) and could replace up to 40% soya meal in chicken diets (ref. 32). However, separation of algae by such a method, is often not appropriate or practicable in many tropical countries at the moment when the more robust techniques are most applicable.

26. A similar approach has also been used based upon the activated sludge technique of waste processing. The dried sludge has been included into artificial diets where it has been shown capable of replacing up to 40% of cotton seed cake/wheat bran mixture commonly used in carp diets (ref. 33).

27. Some assessment of the effectiveness of effluent from a fishpond fed from a high rate stabilization pond as a fertilizer has been made by comparing the results of application to maize with those caused by addition of raw sewage or tap water direct to the maize (ref. 29). The effectiveness of these treatments on growth and yield of maize was directly related to the total nitrogen applied irrespective of its form. Consequently raw sewage was the most effective at promoting growth, followed by the stabilization pond/fish pond effluent, then tap water, with yields ranging from 2,600 Kg grain ha^{-1} to 6,800 Kg ha^{-1} amongst the treatments. Since nitrogen is most frequently the major limiting factor for plant growth in tropical fresh waters (ref. 34, ref. 35) the high nitrogen content of sewage waste in a particularly valuable potential resource in the tropics.

Improvement of effluent quality

28. There are several ways in which the biological community associated with fish ponds can assist in the processing of

sewage waste. For example, algae can remove up to 38% of inorganic nitrogen and 98% phosphorus from cultures owing to rapid incorporation of these components through photosynthesis (ref. 36). In culture, when buffered by an air/CO_2 mixture to prevent the pH exceeding 10, the efficiency of nitrogen removal can be increased to 55% as production increases, although phosphate removal falls. This suggests that the main factor of phosphate removal is actually physical precipitation at the higher pH of the unbuffered medium (ref. 36).

29. The presence of fish in a pond also appears to contribute to the purification of an effluent. Comparison of the final ponds of two series of three sewage-fed ponds run in parallel, one series of which was stocked with silver carp, grass carp and bighead carp, whilst the other series remained as a control without fish, showed considerable differences (ref. 9). Those ponds containing fish had 27% less ammonia and 3% less phosphate than those without, which is partly due to the accumulation of nitrogen and phosphorus in the body of the fish. The BOD_5 in the stocked ponds was 38% lower than in the unstocked ponds, which is partly due to the direct grazing by the fish on the suspended organic particles but also to the promotion of aerobic decomposition during the day by the dense algal blooms which were denser in the stocked ponds, probably due to the stimulating effects of grazing by the herbivores. This also appears to have influenced the algal composition in that blue-green algae predominated in the unstocked ponds, which produced occasional die-backs with the associated deleterious effects on the oxygen relations, whilst small green algae were the principal types in the stocked ponds. Throughout most of the year, the BOD_5 remained well below the widely recommended standard of 30 mg 1^{-1} in the effluent from the stocked ponds whilst in the central ponds it was approached or exceeded on several occasions.

30. One additional particularly significant effect was the progressive reduction in faecal coliform bacteria in both control and stocked series. In the control ponds, for example, that receiving the original sewage effluent showed populations of 500,000 faecal coliforms/100 ml, whilst in the second pond in the chain this declined to 20,000/100 ml and the final effluent from the third pond showed 60/100 ml, which is a safe level for some uses. The stocked series failed to show greater reductions with a final value of 50/100 ml. The reduction of potentially pathogenic bacteria by fish ponds appears to be a consistent but poorly understood phenomenon (ref. 37). It may possibly be due to the alkaline pH promoted by the considerable algal photosynthesis or perhaps due to toxic extracellular secretions by the algae which disinfects the water (ref. 38).

31. The effects outlined above appear to be characteristic of fish or algal ponds in both warm and temperate climates. Passage of sewage affluent through a high rate stabilization pond in Thailand showed mean reductions of COD from 81 to 36.3 mg 1^{-1}, ammonia from 7.64 mg 1^{-1} to 1.1 mg 1^{-1} and a general reduction in nitrogen in the filtered effluent (ref. 10). A small reduction in phosphate was found but generally pond systems seem

most effective at removing nitrogen than phosphorus. In a similar high rate pond in Israel, the effluent showed a reduction in both BOD_5 and COD although the actual amount of suspended matter increased owing to the high density of the algae (ref. 3). All of these factors showed the most marked fall after extraction of the algae by flocculation and flotation and without some form of treatment, ponds with very high densities of algae may not produce effluent of acceptable quality unless the algae are dealt with in some way. For example, in this high rate pond the BOD_5 of the effluent, although lower than the inflow, was still 10^6 mg 1^{-1}, well above the most acceptable level of 30 mg 1^{-1} but after algal separation it was reduced to 10 mg 1^{-1}. The most marked reduction in nitrogen and phosphorus also occurred following the separation process.

32. In both of these systems there was a considerable reduction of potentially pathogenic bacteria through the pond, often by several orders of magnitude. For example, in Israel the coliforms in the effluent had been reduced to 3.5×10^5/100 ml from 6×10^7 in the inflow (ref. 3), whilst in Thailand there was a reduction in faecal coliforms from $0.54 - 240 \times 10^6$ in the inflow to $0.7 - 9.2 \times 10^5$ after passing through the high rate pond and finally to $2.0 \times 10^2 - 1.3 \times 10^5$ after the effluent had passed through the fish ponds (ref. 28). The final effluent from the fish ponds did, however, exceed the WHO limit of 10^2 faecal coliforms for re-use of water for edible crop production. Nevertheless, the disinfecting action of both algal and fish ponds was demonstrated. This, together with the other improvements in water quality produced by aquaculture in conjunction with sewage treatment, gives an added dimension to the benefits of the use of sewage waste for useful production.

CONSTRAINTS TO THE USE OF FISH IN WATERS RECEIVING SEWAGE WASTE
Dissolved oxygen

33. Many fish species used for culture are tolerant of low oxygen conditions and will survive concentrations of 1 mg 1^{-1} or less. This, however, is misleading since feeding can be depressed at concentrations as high as 6 mg 1^{-1} in some species, whilst at levels around 3 mg 1^{-1}, growth and metabolism can become directly restricted (ref. 19). It is of considerable importance to consider all environmental conditions not only in terms of those under which the fish will survive but also in terms of those under which growth will not be inhibited. Raw sewage may itself have a very low oxygen content but even after being retained in a pond long enough for algal blooms to be produced, there is still a considerable daily fluctuation in oxygen from very low in the morning to greater than 100% in the early evening, following photosynthesis during the day. For example, the dissolved oxygen in the high rate pond in Thailand varied between 0.1 and 19 mg 1^{-1} over a daily cycle (ref. 10). Nevertheless, it is possible to add substantial quantities of sewage waste to ponds without endangering the system and it is possible to estimate the most appropriate rate of application (ref. 39). There are also a number of simple aeration devices which can be used to

maintain the oxygen concentration of the water (ref. 40).

Ammonia

34. Ammonia can be particularly toxic to fish, especially under the rather alkaline conditions found in warm-water fish ponds when rapid photosynthesis is taking place. This is due to the progressive conversion of the ionised, relatively non-toxic form of NH_4+ to the unionised, highly toxic NH_3 above a pH of 8. Once again sublethal concentrations will inhibit growth and the check in production noted in intensive pond production of fish in Israel may be due to accumulation of metabolites such as ammonia (ref. 41). In the case of the common carp, beyond a concentration of unionised NH_3 of 0.017 $mgN\ l^{-1}$ there is a progressive reduction in growth rate (ref. 42) which would reduce production. This would be equivalent to values of 0.23 $mgN\ l^{-1}$ for total ammonia at pH8 or 0.075N $mg\ l^{-1}$ at pH9 and 30°C. Concentrations recorded in sewage-fed ponds containing fish far exceed these values; for example the high rate pond of Edwards and Sichumpasak (ref. 10) produced an effluent containing 1.1 $mgN\ l^{-1}$ total ammonia which was fed into fish ponds whilst the fish ponds of Henderson (ref. 9) sometimes contained as much as 13 $mgN\ l^{-1}$ under alkaline conditions.

35. In effluents with a high nitrogen content on a well-developed nitrification system high ammonia values can pose considerable problems. Many algae can, however, utilize ammonia directly as their inorganic nitrogen supply which can mitigate the problem.

Toxic trace pollutants

36. A variety of compounds appear in domestic and industrial effluent which are present only in very low concentrations, yet are highly toxic to fish and, at the same time, present a hazard to the consumer. Such compounds include herbicides, pesticides, heavy metals and hydrocarbons from plastics and oil derivatives.

37. To some extent acute effects can be guarded against by directing some of the incoming effluent through a simple bio-assay tank of fish which would then indicate when the water is toxic and should be directed away from the more vulnerable sections of the stock. However, many of these compounds when present in sub-lethal concentrations accumulate in the body of the living fish, either directly from the water or indirectly via food organisms. For example, the common green algae Scenedesmus has been shown to concentrate heavy metals in water 500-30,000 times with the tendency to concentrate zinc being most marked (ref. 43), whilst fish in the same system possessed concentration factors from 4-1600 with up to 65 $mg\ Kg^{-1}$ dry weight being recorded for zinc in the fish muscle from an initial concentration in the medium of 0.45 $my\ l^{-1}$.

38. This can, therefore, be a potential problem depending upon the concentration of these compounds in the water and the quantities absorbed onto particulate organic material. Some hydrocarbons, whilst not directly hazardous, can give the fish an unpleasant taste (ref. 37).

Diseases of fish and man

39. Fish diseases and parasites do not seem to flourish in sewage treatment ponds and are certainly not more common than in clean waters. There is some evidence to suggest that fish diseases, particularly those affecting the exterior, are actually supressed in sewage treatment ponds (ref. 37).

40. Fish do not seem to actively carry the enteric bacteria which are patgogenic to humans, such as <u>Salmonella</u> and <u>Shigella</u> and, as indicated in the previous section, the general conditions in ponds of this type seem to be unsuitable for these. Such organisms may, however, be passed through the gut of a fish to be transmitted by contact or ingestion. In some parts of the world fish from sewage-fed ponds are kept for a short period of time after harvesting in clean water to allow all gut contents to be evacuated, although in other regions they are used direct and have yet to be implicated as a major source of disease transmission, although little direct evidence is available (ref. 37).

Acceptability and marketing

41. In some cases there is public reluctance to accept fish grown in sewage and consequently the fresh product might need to be introduced gradually, perhaps via some intermediate product such as fish meal. A reduction of fish to meal does, however, generally mean retailing at a lower price. Acceptability is, however, often high although the market may also require familiarisation with a new type of fish if the most appropriate species is new to the area.

REFERENCES

1. MOAVE, R., WOHLFARTH, G., SCHROEDER, G. L., HULATA, G. and BARASH, H. Intensive polyculture of fish in freshwater ponds. I. Substitution of expensive feeds by liquid cow manure. Aquaculture 10, 25-43.

2. RAPPAPORT, U. and SARIG, S. The results of manuring on intensive growth fish farming at the Genosar station ponds in 1977. Bamidgeh 30, 1978, 27-36.

3. SHELEF, G., MORAINE, R., MEYDAN, A. and SANDBANK, E. Combined algae production-wastewater treatment and reclamation systems. Symposium on Microbial Energy Conversion, (Eds.) Schlegel, H. G. and Barnea, J. E. Goltz, Gottingen, 1976, 427-442.

4. MARA, D. Sewage treatment in hot climates. Wiley, New York, 1976.

5. ALWM. Final report on Nile Delta fish farm project. Atkins, Land and Water Management, Cambridge, 1980.

6. JFM. Lake Manzala Study, UNDP. James F. McClaren Ltd., Willow Dale, Canada, 1980.

7. FRYER, G. Conservation of the Great Lakes of East Africa: a lesson and a warning. Biological Conservation 4, 1972, 256-262.

8. SCHROEDER, G. and HEPHER, B. Use of agricultural and urban wastes in fish culture. In, 'Advances in Aquaculture', (Eds.) Pillay, T. V. R. and Dill, W. A. Fishing News Books, Farnham, 1979, 478-486.

9. HENDERSON, S. An evaluation of filter feeding fishes, silver and bighead carp, for water quality improvement. In, 'Culture of exotic fishes symposium proceedings', (Eds.) Smitherman, R. O., Shelton, W. L., Grover, J. M. Fish culture section, American Fisheries Society, Auburn, Alabama, 1978, 121-136.

10. EDWARDS, P. and SINCHUMPASAK, O-A. The harvest of micro-algae from the effluent of a sewage fed high rate stabilization pond by Tilapia nilotica. Part 1: Description of the system and the study of the high rate pond. Aquaculture 1981, 23, 83-105.

11. LUND, J. W. G. Investigation into phytosplankton with special reference to water usage. Freshwater Biological Assoc. Occ. Publ. 13, 1981, 64pp.

12. SUFFERN, J. S., ADAMS, S. M., BLAYLOCK, B. G. Growth of monosex hybrid tilapia in the laboratory and sewage oxidation ponds. In, 'Culture of exotic fishes symposium proceedings', (Eds.) Smitherman, R. O., Shelton, W. L., Grover, J. H. Fish Culture Section, American Fisheries Society, Auburn, Alabama, 1978, 65-73.

13. GAIGMER, I. G. and KRAUSE, J. B. Growth rates of mozambique tilapia (Oreochromis mossambicus) and silver carp (Hypophthalmichthys molitrix) without artificial feeding in floating cages in plankton-rich waste water. Aquaculture 1983, 31, 361-367.

14. BEHRENDS, L. L. Recycling livestock wastes via fish culture. Aquaculture Magazine, 1980, 38-39.

15. SHARMA K. P. Multipurpose use of water resources in relation to the inland fisheries of India. Summary report and selected papers presented at the IPEC Workshop on inland fisheries for planners, (Ed.) Petro. T., Manila, The Philippines, 2-6 August 1982. FAO Fish Rep (288) 1983, 150-166.

16. FAO. Freshwater aquaculture development in China. FAO Fish. Tech. Pap. (215), 1983, 125 pp.

17. LIAO, I-C. and CHEN, T-P. Status and prespects of tilapia culture in Taiwan. In, 'Proceedings of the international symposium on tilapia in aquaculture', (Eds.) Fishelson, L. and Yaron, A. Tel Aviv University, Tel Aviv, 1984, 588-598.

18. GALMAN, O. and AVTALION, R. R. A preliminary investigation of the characteristics of red tilapias from the Philippines and Taiwan. In, 'Proceedings of the international symposium on tilipia in aquaculture', (Eds.) Fishelson, L. and Yaron, Z. Tel Aviv University, Tel Aviv, 1984, 291-301.

19. PAYNE, A. I. Physiological and ecological factors in the development of fish culture. Symp. Zool. Soc. Lond. 44, 1979, 383-415.

20. GLOYNA, E. F. Waste stabilization ponds. World Health Organization, Geneva, 1971.

21. OSWALD, W. J. and GOTAAS, M. B. Photosynthesis in sewage treatment. Trans. Am. Soc. Civil Engrs. 122, 1957, 73-105.

22. OSWALD, W. J., GOLUEKE, C. G. and HORNING, D. O. Closed ecological systems. Proc. Am. Soc. Civil Eng. J. San. Eng. Div. 91, 1965, 23-24.

23. SHELEF, G. and HALPERIN, R. Wastewater nutrients and algal growth potential. In, 'Developments in water quality research', (Ed.) Shuval, H. I. Ann-Arbor-London, Ann-Arbor-Humphry Sci. Pub., 1970, 211-228.
24. BAUMERT,H.and UMLMANN,D. Theory of the upper limit of phytoplankton production per unit area in natural waters. Int. Revue ges Hydrobiol. 68, 1983, 763-783.
25. McGARRY, M. G. and TONGKASAME, C. Water reclamation and algae harvesting. J. Water Pollut. Control. Fed. 43, 1971, 824-835.
26. EDWARDS, P. A review of recyling organic wastes into fish, with emphasis on the tropics. Aquaculture 21, 1980, 261-279.
27. EDWARDS, P. SINCHUMPASAK, O. and TABUCANON, M. The harvest of microalgae from the effluent of a sewage fed high rate stabilization pond by Tilapia nilotica. Part 2: Studies of the fish ponds. Aquaculture 23. 1981a, 107-147.
28. EDWARDS, P., SINCHUMPASAK, O., LABHSETWAR, V. K. and TABUCHANON, M. The harvest of microalgae from the effluent of a sewage fed high rate stabilization pond by Tilapia nilotica. Part 3: Maize cultivation experiment, bacteriological studies and economic assessment. Aquaculture 23, 1981b, 149-170.
29. LINCOLN, E. P., HILL, D. T. and NORDSTEDT, R. A. 1978. Harvesting algae from lagoon effluent. Agric. Eng. 59, 1978, 16-18.
30. MESKE, V. and PRUSS, H. D. Fish meal free food on the basis of algae powder. Adv. Anim. Physiol. Amin. Nutrit. 18, 1977, 71-81.
31. HEPHER, B., SANDBANK, E., and SHELEF, G. Fish feeding experiments with wastewater grown algae. In, 'Combined systems for algal wastewater treatment and reclamation and protein production. 2nd Progress Report, Technician. Haifa, 1975.
32. MUKADI, S. and BERK, Z. Feeding experiments of chicken with sewage grown algae. 2nd Progress Report. Technion, Haifa, 1975.
33. ANWAR, A., ISHAK, M. M. EL-ZEINY, M. and HASSANEN, G. D. I. Activated sludge as a replacement for bran-cotton seed meal mixture for carp, Cyprinus carpio L. Aquaculture 28, 1982, 321-325.
34. ZARET, T. M., DEVOL, A. H. and DOS-SANTOS, A. Nutrient addition experiments in Lago Jacartinga, Central Amazon, Brazil. Verh Internat. Verein Limnol. 21, 1981, 256-259,
35. MOSS, B. Limitation of algae growth in some Central African waters. Limnol. Oceanogr. 14, 1969, 591-601.
36. KAWASAKI, L. Y., TARIFENO-SILVA, E., YU, D. P., GORDON, M. S. and CHAPMAN, D. J. Aquacultural approaches to recycling of dissolved nutrients in secondarily treated domestic waste waters I. Nutrient uptake and release by artificial food chains. Water Res. 16, 1982, 16-37.
37. ALLEN, G. H. and HEPHER, B. Recyling of wastes through aquaculture, and constrains to a wider application. In, 'Advances in Aquaculture', (Eds.) Pillay, T. V. R. and Dill, W. A. Fishing News Books, Farnham, 1979, 478-486.
38. DAVIS, N. E. A manual of wastewater operations. 4th Edition. Lancaster Press, Lancaster, Pennsylvania, 1971.

39. SCHROEDER, G. Some effects of stocking fish in waste treatment ponds. Water Res., 1975, 591-593.

40. RAPPAPORT, U., SARIG, S. and MAREK, M. Results of tests of various aeration systems on the oxygen regime in the Ginosar experimental ponds and growth of fish there in 1975. Bamidgeh 28, 1976, 35-49.

41. RAPPAPORT, U. and SARIG, S. The results of tests in intensive growth of fish at the Genosar (Israel) station ponds in 1974. Bamidgeh 27, 1975, 75-82.

42. JACKSON, W. T. The influence of high population densities on the growth of Cyprinus carpio (Linn.) and larvae of Xenopus laevis (Dandin) with particular reference to chemical factors. PhD Thesis, Coventry (Lanchester) Polytechnic, Coventry, 1983.

43. TARIFENO-SILVA, E., KAWASAKI, L. Y., YU, D. P., GORDON, M. S. and CHAPMAN, D. J. Aquacultural approaches to recycling of dissolved nutrients in secondarily treated domestic wastewaters - III. Uptake of dissolved heavy metals by artificial food chains. Water Res. 16, 1982, 59-65.

10 A Dutch example and continental practice

Ir H. M. J. SCHELTINGA, HonFIWPC, Ministry of Housing, Physical Planning and Environment, Arnhem, The Netherlands

SYNOPSIS. In constructing the new Flevo polders in the former Zuiderzee, lakes remained between the "old land" and the polder dykes. These lakes are intensively used for recreation, but also for many other functions such as recharging the polder-canals for level control in summertime. The water quality in the lakes is severely influenced by many direct and indirect effluent discharges. Phosphorus removal and effluent disinfection are therefore standard procedures in sewage treatment. Improvement of the water quality is also effected by deliberate water quantity management. Results are compared with the original situation.

INTRODUCTION. To survey the European situation as to the re-use of effluents for amenity purposes is a rather difficult task. Hardly any data are available on the subject. In defining the scope of this approach it should be stated that only those situations will be discussed in which bodies of water with the primary functions "recreation and/or sportfishing" receive effluent in relatively large quantities. Difficulties in meeting water quality criteria in such situations are to be expected especially in stagnant waters. Recreation projects based upon the presence of surface waters, are rather often situated in (artificial) lakes, constructed for balancing flow in small rivers.The consequences of combining the recreational function with that of reception of effluents may cause problems. As an example, the rather famous situation in the Netherlands: the Veluwe-lake will be described in some detail.
Lake Veluwe, which is directly connected with the Dronten lake, was created about 1956, when the dykes for the polder Flevoland were constructed (Fig. 1).

Reuse of sewage effluent. Thomas Telford Ltd, London, 1984

173

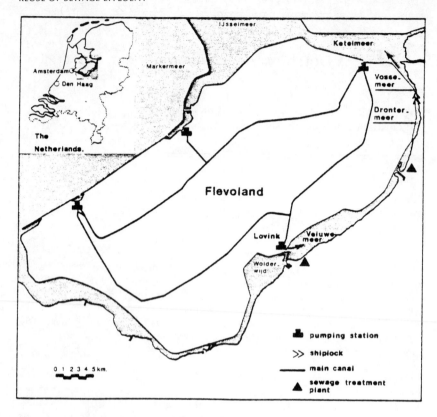

Fig. 1   Location of Lake Veluwe (Veluwemeer)
         and Lake Dronten (Drontermeer).

HYDROLOGICAL AND ENVIRONMENTAL FACTS OF VELUWE LAKE
In reclaiming the Zuiderzee, the first and most important
part was the construction of the Great Barrier Dam. This Dam
creating the IJssel-lake was completed in 1932. Within five
years, the chloride-ion concentration in the water dropped to
around 160 mg/l, the same as that of the river Rhine in those
years. In the IJssel-lake two polders were constructed without a peripheral lake between the polder and the "old
land". It was found however that drying-up symptoms were
appearing in the "old" land next to the North-East polder.
The same symptoms were to be expected and to a greater degree
when the third polder Flevoland was created and this was the
reason for creating peripheral lakes between the "old" land
and the polder. These lakes vary in width, the determining
factor being the geohydrological conditions of the particular
area. The wider lakes mainly occur where the transmissivity
of the subsoil is great. A narrow lake suffices when the
transmissivity is low. Three locks were constructed to enable

174

the level of the lakes to differ from that of Lake IJssel. This led to the creation of two compartments. The peripheral lakes can also be used to limit the inflow of brackish or salt groundwater. This inflow is a typical Dutch process which has increased since the seventeenth century through human intervention, notably as a result of the reclamation of land, known as polders. The flows of groundwater to these low-lying, reclaimed areas cause brackish water to well up and this in turn increases the salinity of the surface water.The deeper the polder, the greater the force driving the groundwater flow and thus the higher the salt-level. Possible environmental problems were of little interest in those days. Consequently, no one was concerned about possible changes in the bacteriological or chemical quality of the water as a result of the total project. In the late sixties there was a hesistant start to the regular quality control of the water, also due to the fact that recreation was increasing while water quality was decreasing. Then the Pollution of Surface Waters Act was passed in December 1970. This gave a major boost to concern about water quality. The setting up of regional working parties on the prevention of water pollution was such a step along the road. Two such working parties were instituted for the Lake IJssel region by the Minister of Transport and Public Works: the Peripheral Lakes Working Party came into being in April 1974 and the Lake IJssel Working Party in June 1975. Their terms of reference were as follows: 'to study the question of cleaning up the peripheral lakes and Lake IJssel (including listing the existing effluent discharges and forming some picture of the nature and scale of the sources of pollution and how they are expected to develop in the future) and, in the light of the data and understanding acquired, to indicate guidelines required, to indicate guidelines for the most efficient means of cleaning up the peripheral lakes and Lake IJssel'. In studying the peripheral lakes, the working party discovered that it was difficult, if not virtually impossible, to compile the water and materials balances owing to a lack of adequate data. The availability of data on waste and effluent discharges and on water quality on the old land, in particular, left much to be desired. Data on the quality of the water in the peripheral lakes began to be collected once a fortnight at 19 sites in 1972. At the same time, data were collected on the water quantities from the different sources. This finally resulted in the figures presented in table 1.

Table 1. Inflow in Veluwe-lake in %

Rain	10%
Small rivers ("old land")	20%
Underground flow	15%
Pumping stations (polder)	50%
Effluents	5%

These figures are based upon average data over a year (since 1979). In the summertime, the most critical period for water quality, effluents will constitute about 25% of the total inflow.

FUNCTIONS OF THE LAKE

The most important functions of the Veluwe lake are: recreation, fishing both commercial and for sport, reception of surplus water from the "old land" and the polder Flevoland and finally shipping, recreational as well as commercial.

Based upon an extensive report of some years ago and shipping data of 1981 (ref 1,2) some figures are presented to illustrate the importance of these functions. The lake is about 17 km long, having a width of ca. 800 m in the North-Eastern part of 7 km and a width of 2500 m in the other part.

On this relatively small area many people and many ships are seeking a space. About 20.000 holiday makers are present on a good summer day, together with 3000 sailing boats. Within ten years it is expected that these figures will be 50% greater. At this moment there is harbour capacity for about 5000 boats. It is expected that the capacity will be extended to 9000 in the year 2000. Commercial shipping, using the lake all over the year, totals 2000 passages per year and is thus of minor importance. About 20 licences for commercial fishing are supplied per year. Estimating the number of people practising sport fishing is very difficult. All together, in all the peripheral lakes, perhaps 40.000 persons may fish regularly. Ten percent of this total number will use the Veluwe lake. On suitable days, a few thousand fishermen (someusing rowing boats etc.) can be expected.

WATER QUALITY CRITERIA.

For the function water for (commercial) fishing and for bathing the formal European guidelines, incorporated in the Pollution of Surface Waters Act 1970, apply.

These directives are in use in all the EC countries. They play an important role in the water quality planning. For those functions that are not covered by "Brussels" directives, often national guidelines apply. Requirements for a "basic" or "minimum" quality are known in many countries or regions. They apply mostly for general purposes such as non-contact recreation, water for agricultural use, for sport-fishing etc. Some attention will be given here to basic objectives. The determination of water quality objectives for all of the beneficial uses of waters is a task beyong the scope of this paper. However, in arriving at water quality objectives all beneficial uses should, if possible, be taken into account and the quality objectives must be designed to meet the most stringent use requirements. This use often will be one assiociated with fisheries or recreation. Ecological quality objectives should meet the needs of all individual aquatic ecosystems and should, therefore, not be applied nationally but rather regionally, in consultation with local authorities. There is always a danger that water quality objectives will be looked upon as ideals which may be sought

176

after, but which need not be reached in practice. It is important that they be regarded as minimal standards and that, whenever possible, superior standards should be the aim.
There is also the possibility that, once prescribed, quality objectives become fixed and unchangeable. Quality objectives should be recognised as values based on our present state of knowledge. There should be freedom to revise them as necessary to meet higher demands of environmental quality, to adapt to new knowledge of dosis/effect relationships and to technical progress in industrial processes and effluent treatment. In choosing the parameters to be dealt with as a first priority two groups of criteria can be used in characterising a basic quality:
- the basic characteristics of water quality such as DO, BOD, nutrients, pH, suspended solids, temperature, and
- the toxic or harmful pollutants which are to arise most commonly: cadmium, chromium, copper, mercury, other heavy metals and organic micropolutants.

Not unusual are the following figures for these parameters:

Dissolved oxygen	> 5 g O_2/m^3 or $> 50\%$ saturation
BOD - 5	< 5 g O_2/m^3
Nutrients	NH_4-N < 1 g/m^3, total P $< 0,2$ g/m^3
pH	$6.5 - 9.0$
Temperature	< 25 °C or maximum 3 °C over natural temperature

Maximum acceptable levels for heavy metals are mostly expressed in total concentration. "Black listed" substances as mercury and cadmium show levels of only 0.5 and 2.5 mg/m^3. The other heavy metals lay in the order of 50 mg/m^3.
There is a tendency to define the standards for heavy metals in a soluble part and a part absorbed on suspended solids. It seems to be a rather logical approach as the main part of these substances is absorbed. The concentration in a water largely depends upon the content of suspended solids. In stagnant water the bottom of a canal or lake often shows very high levels of metals, due to deposition of suspended solids. The disposal of dredged sludge may become very problematic because of the quality. The "supernatant" water however has very low concentrations of heavy metals. The first compartment of the IJssel lake shows this phenomenon very clearly. This Ketel-lake works like a sedimentation tank in a sewage treatment plant for the suspended solids in the Rhine river water.

WATER QUALITY, 1970. In 1970 the two main sewage works were in operation, without tertiary treatment. The small rivers were heavily polluted by discharges of manure, direct and indirect by run-off.
Chlorination of effluents started in May 1970. Table 2 gives

177

the MPN-figures for five sampling points on both banks of the
lake opposite (A) and near (E) the effluent discharge point
of the Harderwijk plant.

Table 2. MPN fecal coli per ml, 1970. (ref 3)

Point	13 May	19 May	25 May	16 June
A	93	9	0.9	0.5
B	430	4	2	0.7
C	930	2	2	2
D	930	15	9	0.8
E	70.000	43.000	-	13

The original situation as shown by the 13 of May figures was
completely unsatisfactory. Eutrophication was in 1970 a real
handicap for the use of the Veluwe lake for recreational purposes. Retention times in summer are about half a year or
more, depending on rainfall thus promoting algae bloom. In
these days the input of phosphor per year was ca. 100 tons of
P. About 75% originated from effluents of the two sewage
purification plants. (ref 4)
It was then estimated that a loading of 1 P per m^2 and year
was acceptable for this type of lake, what means a loading of
40 ton of P. By P-elimination from the effluents this goal
could be reached. Both plants now use since some years a
chemical simultaneous flocculation of phosphorus, showing an
overall efficiency of 90% at least.
Another problem was the deposition of phosphorus in the
bottomsludge. The benificial effects of this effluent
P-elimination were counteracted by the fact that much P had
accumulated in the bottomsludge. Experiments showed that,
favoured by anaerobic conditions down below, recycling of P
into the supernatant water takes place. The 1970 quality of
the water can be described as extremely poor. Transparency of
ca. 20 centimeters in summer-time, a green colour due to
massive algae growth and a very unbalanced dissolved oxygen
situation.
Data for chlorophyl-a concentrations in that time are not
available but they must have reached far over 200 mg/m^3.
Under such circumstances the fish population is threatened
each year again. For swimming and other recreational functions the quality of the water in the holiday season was
hardly or not sufficient. Several measurements were taken to
improve the situation as the economical importance of the
recreation in this regions was growing steadily.

IMPROVEMENTS. The authority responsible for the water quality
in the lake, the Directie Zuiderzeewerken from
Rijkswaterstaat, realised several measures to improve the
situation. As to the hygienic quality of the water, des-
infection of effluents was enforced as early as 1970, for the
two large purification plants. Further research was started
to determine the importance of pollution caused by pleasure
178

boats. It took a rather long time before the results of this study were available. Again, as with the figures in the water balance, no data were available anywhere. (ref. 5)

The main influence on the hygienic quality of the water is excercised by boats in harbours, using pump toilets. The installation of sanitary facilities in the harbours, together with the issue of a prohibition to use pump toilets during stay in the harbour was effectuated. Similar facilities were also made available on other recreational concentration points, used for swimming.

The abatement of the poor waterquality due to eutrophication is far more complicated. First of all the permit for discharge of effluents was changed in such a way that P-elimination had to take place with an efficiency of 90%. Until recently this form of tertiary treatment was subsidised by Rijkswaterstaat in the two plants.

Another very effective procedure was introduced in 1979. In wintertime the surplus water from the polder was by preference pumped into the Veluwe lake instead of in the big IJssellake. This rainwater, low in phosphor, diluted the rather eutrooph water that the Veluwe lake contains at the end of the summer. A small part of the P accumulated in the sludge may, by this procedure, be taken out of the lake simultaneously. "Cleaning" the bottom of the lake, as has been done in small research objects, has not been effectuated yet. It is of course a very expensive way of eliminating an important source of phosphor in the lake. There is some hope that the "winter-dilution" proces will have the same effect, but after a much longer period.

WATER QUALITY, 1982. The quality of the water in the lakes under supervision of the Directie Zuiderzeewerken, is under regular control at many sampling points. On behalf of holiday-makers general information on leaflets is given about the bacteriological quality of the water in bathing areas. In general the quality now is good to at least acceptable, according to the EEC directive. The information of the public at the beginning of the season is based upon last years data. As there are no changes in the situation the last few years, this procedure seems to be justified.

The ideal waterquality for the Veluwe lake can be described with the following figures (ref 6): total P 0.05 - 0.10 g/m^3 together with chlorophyl-a 25 - 50 g/m^3 and a transparency of 1.50 - 0.75 meter. These levels are supposed to be a realistic goal under the following conditions: all effluents being dephosphated, normal manurial practice in agricultural regions on the old land and continuous use of the "winter dilution". An extremely difficult subject is the abatement of dumping practice of pig and calves manure, causing not only eutrophication of the small rivers discharging into the lake, but also of the groundwater, reaching as underground flow the the lake also.

The phosphor loading of the Veluwe lake is now about 0.8 g P/m^2 year, the target in the early seventies. The chlorophyl-a content of the water now lies in the range of 100 mg/m^3 but in 1975 and later it used to be 200 - 300 mg. There is a distinct improvement. The same holds for the dissolved oxygen, the extreme values in the season as well as in the day-night rithm disappeared to a great extent. The transparency, in the seventies about 20 cm, doubled but is still too low, as the target is at least 75 cm.

In testing the analytical results with the standards of the directives for the different functions it shows that all the bathing areas are all right, except for the parameter transparency.

The requirements of the directive for fish life (ciprinide fish) in general are met but in a few locations P-levels are too high. The same holds for the pH. The standards for the national basic quality are reached but mainly not for phosphor. Alltogether there is a distinct improvement in quality since the year 1970. It should be realised that in this period the general situation changed unfavourable in the sense of growth of the total pollution load produced in the region as well as of the number of recreation facilities and the people visiting them, with or without boat.

CONTINENTAL PRACTICE

Similar situations as described in this paper do occur in a number of other regions in the Netherlands. The surface waters in our delta tend to be rich in nutrients, so eutrophication is a serious threat for the recreational functions of many stagnant waters.

The same holds for lakes in other parts of Europe. In Germany several lakes in the densely populated districts of Ruhr and Emscher receive special attention of the "Wasserverbände" in order to protect their functions for drinking water production and for recreational purposes. The same holds for many Bavarian lakes, where a very good transparency, some sort of touristic attraction, is protected by high standard biological treatment of sewage, including tertiary treatment (P-elimination and denitrification).

A very recent example of quality management is the Tegern-lake in West-Berlin. Here not only P-elimination of purified effluents but also of the surface waters feeding the lake is effectuated.

In Switserland the same type of treatment is given in such cases. Often the total load of waste water is transported by a peripheral sewerage system, to a purification plant situated at the bottom of the lake, or even on its outgoing river. This procedure is also known in the French Alpine district for lake management.

This strategy to protect recreational waters, will give the safest solution but may be rather expensive if very long distance transport will be necessary.

Full re-use of effluents for amenity purposes in the most direct sense, is in my knowledge not effectuated anywhere in

180

Europe. Such a situation is hardly to be expected in Westeuropean regions but rather in arid climates. On the other hand in very many cases the recreational function of a water must be combined with that of reception of effluent. It is hoped that the description of the "history" of the Veluwe lake with a very unusual combination of factors, can serve as an example in less unusual situations elsewhere.

ACKNOWLEDGEMENTS

Thanks are due to Ir. M. Snijdelaar, Directie Zuiderzeewerken, for his invaluable help, direct as well as indirect, in the preparation of the paper. I also like to thank my colleague Ir. S.H. Hosper from Rijkswaterstaat, Inland Waters Research Division, for his advises and corrections.

REFERENCES
1. Report: Ontwikkelingsvisie recreatief gebruik randmeren. Directie Zuiderzeewerken, Lelystad, 1977.
2. Scheepvaarttellingen IJsselmeergebied. Directie Zuiderzeewerken, Lelystad, 1983.
3. Report: Waterkwaliteit in de randmeren langs Flevoland. Werkgroep Coördinatie kwaliteitsonderzoek randmeren, 's-Gravenhage, 1972.
4. Report: Subwerkgroep 3, discharges. Werkgroep Sanering Randmeren. Rijksdienst voor de IJsselmeerpolders, Lelystad, 1877.
5. Report of the Werkgroep verontreiniging van recreatiewateren door de pleziervaart. Staatsuitgeverij, 's-Gravenhage, 1981.
6. Waterkwaliteitsplan IJsselmeer, in preparation.

11 Guidelines for evaluating recreational water reuse

J. F. CARUSO, Department of Public Works, Nassau County, New York, and
R. J. AVENDT, PhD, PE, Environmental Engineering, Greenhorne & O'Mara,
Inc., USA

SYNOPSIS. Although recreational water reuse applications account for only a small quantity of water reused in the United States today, in certain communities this type of reuse plays a significant role. A critical assessment of the planning and management concerns; water quality requirements; potential health effects and social, legal and institutional factors must be made for each proposed facility. The guidelines for evaluating recreational water reuse projects are presented together with the results of several successful projects.

INTRODUCTION

1. The Nassau County, New York, Department of Public Works is currently developing a comprehensive ground water management program. This program is the result of several factors which have been extensively evaluated by the County. The continued use of a limited ground water supply to meet increasing potable water demands will result in a 20 million gallon per day deficit by the year 1990. Recent changes in regulatory policies at the Federal and State levels will require the County to assume increased administrative and financial responsibilities for ground water management. Compliance with environmental regulations is becoming more complex and affects the mangement of water supplies, wastewater, solid and hazardous wastes disposal. The financial commitment of the County will become more pronounced with further reductions in public works grant assistance and continuing increases in labor, energy, and material costs.

2. Two areas of immediate critical concern are ground water protection and conservation. The increasing potable water demands of Nassau County are provided by a limited ground water supply. Several recent instances of contamination by industrial wastes and landfill leachates have demonstrated the need for prudent management of this resource. One method of resource management being investigated is wastewater reuse.

3. Wastewater reuse applications may be classified according to four (4) distinct types for purposes of this paper i.e., ground water recharge, agricultural reuse, industrial applications and recreational use. This paper will discuss the investigations the County has undertaken to determine the feasibility, environmental impact and public acceptance of recreational water reuse.

4. Reclaimed water fits naturally into recreational applications such as the development and maintenance of recreational and scenic parklands. The County enjoys an extensive network of bicycle paths and walking trails adjacent to existing stream beds. Augmentation of these streams with reclaimed water has been recommended as a form of "non-body contact" type of recreational water reuse. Other types of non-body contact water reuse are the use of ponds or impoundments for containing reclaimed water for boating or sport fishing. "Primary body contact" activities include swimming and skiing.

5. The public acceptance of utilizing reclaimed wastewater for recreational purposes will vary regionally as alternative water supplies dwindle. Overcrowding of the limited water related recreational sites will require new facilities to be constructed. The use of reclaimed wastewater to create or maintain these recreational sites requires prudent planning. The benefits of water reuse for recreational purposes need to be evaluated on a local perspective consistant with the local impacts.

6. Any consideration of recreational water reuse should address the planning and management concerns; requirements for water quality; potential health effects and social, legal and institutional factors. Existing case histories of recreational water reuse should be evaluated to determine the benefits and mitigation of adverse impacts.

PLANNING AND MANAGEMENT

7. The planning and management of recreational water reuse facilities requires that the concerns normally associated with both wastewater reclamation and water supply be evaluated. The overall procedures for developing wastewater reuse programs are unique for each potential application. Several computer programs are available to speed up and simplify the calculations necessary for determining the wastewater reclamation process scheme, the size and cost of the reclaimed water delivery and containment systems and present worth determinations.

8. The management structure will also cross traditional boundaries between freshwater use and wastewater treatment.

The operational procedures must provide fail-safe performance and minimize the potential adverse conditions that can arise in the course of implementing a wastewater reuse system, especially with respect to public health.

9. Procedures for planning a recreational water reuse project essentially follow an outline similiar to that for planning water or wastewater facilities, with the exception that an initial feasibility assessment is required. Following the feasibility assessment, an engineering project report is prepared, depending on whether or not the reclaimed water is capable of meeting the criteria established for potential recreational reuse facilities. After a recommended alternative is selected from the feasibility report, the remaining activities are accomplished; predesign, design, construction start up and operation.

10. Feasibility assessment. Initially, a feasibility assessment determines the potential recreational uses of reclaimed wastewater within a study area. The water quality and quantity requirements of existing potential uses are also determined. The major potential uses for recreational reuse are landscape irrigation, recreational impoundments and streamflow augmentation. The recreational reuse planner first must assemble a list of potential users and their water quantity and quality requirements. An initial listing actually can be obtained from either parks and recreational departments or municipal agencies. Pertinent data are the amount and types of recreational water uses within the study area, current costs associated with water supplies, incentives for water reuse, and specific constraints associated with reclaimed wastewater use. If local park and recreational departments and municipal agencies are amenable to recreational wastewater reuse, then their cooperation is invaluable in obtaining public support.

11. After potential recreational reuse alternatives are identified, they should be screened and categorized into groups based on water quality requirements, volume of water received, economic considerations and so on. The water quality criteria are used to determine in advance whether a potential recreational use is feasible given the costs and other constraints. Water volume is important because of the economics of scale in large water reclamation process schemes. Site specific concerns regarding public health, legal and institutional factors should also be evaluated to determine the project feasibility.

12. The general nature of a feasibility assessment may be illustrated by the streamflow augmentation studies performed by Nassau County. These studies reported that

Table 1. Wastewater reclamation/reuse project outline

I. Study area characteristics
   A. Topography:
      1. Wholesale and retail water purveyor boundaries
      2. Wastewater agency boundaries
      3. Hydrologic features
   B. Water supply system
      1. Groundwater and surfacewater sources
      2. Treatment plants, pipelines, reservoirs
   C. Wastewater disposal system
      1. Interceptors, treatment plants
      2. Disposal facilities
   D. Reclaimed wastewater system
      1. Existing facilities
      2. Potential new users; quantity and quality requirements
II. Water supply characteristics
   A. Facilities (supply, treatment, transmission, storage)
   B. Costs and revenues
   C. Future demands: quality and quantity
III. Wastewater treatment and disposal
   A. Facilities (capacities, processes, required expansions)
   B. Effluent quality and quantity (flow variation)
   C. Existing reuse (users, quantities, contractual and pricing arrangements)
IV. Wastewater reclamation/reuse potential
   A. Market assessment
      1. Users (quality and quantity requirements, onsite costs, incentives, reliability needs, disposal methods)
      2. Purveyors (service area, pricing structure, legal constraints to reuse)
      3. Suppliers (facility capacity and processes)
   B. Preliminary feasibility analysis
      1. Technical
      2. Economic
V. Alternative analysis, including environmental impacts
   A. Planning and design criteria (removal efficiencies, costs, peaking factors, discount rate)
   B. Treatment alternatives (processes, pretreatment versus internal treatment)
   C. Distribution alternatives (system capacity, routings, storage locations)
   D. Alternative costs (capital, O&M, resources—especially energy)
   E. Environmental impacts of all alternatives (pollution control, water quality of receiving body and groundwater, solids disposal, health effects)
   F. Comparison with other alternatives (freshwater development, conservation, no project)
VI. Recommended plan
   A. Facilities description (treatment, distribution network)
   B. Cost estimates projected over facility lifetime
   C. Extent of service (users, quantity and quality provided, reliability, cost to user)
   D. Implementation plan (actions by users, purveyors, and suppliers; scheduling, permits, commitments)
   E. Operating plan (responsible agency, source of funding)
VII. Financial plan
   A. Allocation and timing of costs (users, purveyors, wastewater dischargers)
   B. Sources and timing of revenues (grants, construction loans, user charges)
   C. Pricing strategies vis-à-vis freshwater resources (sunk costs and bonded indebtedness)
   D. Sensitivity analysis of financial projections

under natural conditions, streams on Long Island derive about 95 percent of their discharge from the ground water reservoir. Heavy ground water withdrawals coupled with pronounced reduction in streamflow. At times, the upper reaches of some of the streams are dry, and as the water table continues to drop, more of the stream channels will be left dry. A stream channel which was once draining water from the ground water reservoir could be used as an area of water spreading to recharge the ground water body. In addition, augmentation of streamflow at the upper reaches of a stream would help maintain higher flows throughout the downstream length of the stream, thus providing for aesthetic and recreational benefits as well as for recharge benefits.

13. Project report. Following the feasibility report, an engineering project report should be prepared detailing the study area; water supply and wastewater characteristics; treatment requirements for discharge and for reuse; wastewater reclamation potential; recommended alternatives; and economic, financial and institutional viability. If further development seems likely, then the engineering report may be expanded to serve as the design and operational reports for the proposed recreational reuse project. An outline of the major items to be discussed in the project report is presented in Table 1. This outline will serve to identify other means of reuse that may be implemented in a study area.

14. Completion of the feasibility assessment and project report with its recommended alternative represents less than half of the overall project process. The remaining tasks to be completed are the design and construction activities, cost distribution, management and operation. One principle output from the project report should be the recommended institution responsible for the reuse project. This institutional arrangement may be private or a government agency. The responsible agency now must investigate the mechanism for financing design and construction. By the time the design is complete the agency should have signed agreements between reclaimed water supplier and the reuse facilities, discharge permits and operating agreements. Necessary reclaimed water use ordinances and operating manuals should also be finalized.

15. At present in Nassau County, recreational water reuse by streamflow augmentation is in the project report phase. The feasibility studies indicated that the most attractive feature of reuse through streamflow augmentation is that little if any land other than the natural stream channel is required. Because land is at a premium in Nassau County, this method of recreational reuse is thought to

be worthy of further detailed investigation, possibly as a demonstration program. The benefits identified from the reports completed to date include: the propagation of more desirable species of fauna and flora; improvements to stream quality; the channel will remain stable and capable of handling discharge peaks with less flooding and erosion; water which would otherwise be lost is restored to the groundwater reservoir; undesirable contaminants will be flushed from parts of the aquifer system and adjoining parklands will be aesthetically improved. Detailed cost analyses and implementation schedules remain to be finalized. The required water quality for this method of reuse has been tentatively set at approximately twice the state drinking water standards. A total of 34 million gallons per day of reclaimed water would be discharged to the headwaters of various streams in the County.

16. <u>Management and operation</u>. A problem common to all water reuse projects is that reclaimed water is not a cheap water supply. The high quality requirements associated with recreational water reuse dictates the principal costs. Where the cost of reclaimed water exceeds the price that can be charged or recovered locally, the recreational benefits on a regional, state or federal level need to be identified, quantified and recaptured. This compensation or subsidy recaptures the indirect benefits that are associated with recreational water reuse. Involving all the indirect bene- ficiaries early in the planning process, and providing for their presence in the ongoing management structure, makes it easier to recapture indirect benefits from a local reuse project.

17. A number of institutional arrangements may be considered for financing, administrating and operating reuse projects. A regional approach to recreational reuse development is likely to provide efficiency in planning, funding and con- struction through the use of common priorities, criteria and objectives for overall water resource management. Financial support may be facilitated by a regional or state agency lead. It is extremely doubtful that this arrangement would be directly involved in ongoing operation and mainten- ance. Administration and assignment of liability for the recreational water reuse project will be subject to the existing laws and regulatory agency guidelines. Financial planning and user charges must accommodate the preceived versus actual benefits. Subsidization beyond user fees commonly has been in the form of ad valorem taxation at local or regional levels, and grants or long term loans with exceptionally favorable terms in the case of state and federally funded projects.

18. Because the reclaimed water may be of poorer quality, both chemically and bacteriologically, then the freshwater it replaces, care must be taken in establishing the operational guidelines to assure water quality standards and protect public health.

WATER QUALITY CRITERIA
19. Certain water quality parameters are of special significance for wastewater intended for reuse in recreational applications.

20. <u>Contact recreational water</u>. The standards for contact recreational water quality are higher than for noncontact reuse applications. In fact, it is common for effluents intended for body contact recreational uses to receive tertiary treatment. The following conditions for water used for primary contact recreation have been taken from the literature:

a. Water must be aesthetically enjoyable.
b. It must contain no substances that are toxic on ingestion or irritating to the eyes or skin of human beings.
c. It must be reasonably free from pathogenic organisms. Care must be exercised in using coliform as indicator organisms, as many of the diseases related to swimming in polluted water are not enteric.
d. Other concerns are temperature, pH, chemical composition, and aquatic growths or clarity of the water. Use of an effluent with BOD and suspended solids values less than 20 mg/L will reportedly achieve the desired goal.
e. Clarity is important for recreational waters for a variety of reasons, which include safety, visual appeal, and recreational enjoyment.
f. A fecal coliform standard is intended to provide for the enjoyment of limited contact users in relative safety. Such uses include boating, fishing, and other non-body immersion activities incident to shoreline usage; whole-body immersion recreation refers to those activities such as bathing, swimming and waterskiing in which there is prolonged, intimate contact with the water and considerable risk of ingesting quantities sufficient to pose a significant health hazard.
g. In bathing and swimming waters, the acceptable range of pH is 6.5 to 8.3 because the lacrimal fluid of human eye is approximately 7, and a deviation of

pH from the norm may result in eye irritation. Well-buffered water may increase the range to 6.0-to-9.0.

h. Chemical characteristics for primary contact water should indicate the water to be non-toxic and non-irritating to the skin, mucous membranes, or ears of humans.

21. Limited body contact, recreational water. Representative water quality criteria for limited body contact or secondary contact (canoeing, fishing, and so on) are less restrictive than those for primary contact. Generally, a well-oxidized secondary effluent is satisfactory. Coliform counts up to 50 000/100 mL have been used. Coliform numbers are highly variable in different applications, with no generally agreed on value.

22. Non-contact recreational water. Certain water quality criteria are established to protect the biota and provide conditions conductive to the growth and propagation of fish:

a. Except where caused by natural conditions, water temperatures of swimming and fishing waters in excess of 30° C (86° F) are not acceptable. Excessively high temperatures are damaging to aquatic biota.

b. Recreational lakes that are filled and maintained with wastewater effluents are characterized by a high eutrophication potential; the possibility of excessive algae growth is caused by nutrients present in treated wastewaters. Therefore, recreational use should require a reduction in the phosphorus and nitrogen content of the water.

c. Shellfish species available for harvest should be fit for human consumption. Therefore, consideration must be given to factors affecting the farming of shellfish, including microbiological quality, pesticides, marine biotoxins, trace metals, and radionuclides.

CASE STUDIES
23. Significant, successful case histories of recreational reuse applications are discussed below:

24. Lancaster, Calif. Since 1974, the sanitation districts of Los Angeles County have sold renovated wastewater to the county for use in a chain of three recreational lakes at Lancaster. The lakes have a capacity of 300 ML (80 mil. gal) and serve as the focal point for the county's 22 ha (56-acre) Apollo Park. The park features fishing, boating, and picnic areas.

25. Lubbock, Tex.-Yellowhouse Canyon Lakes Park. Lubbock residents and visitors can now enjoy the Yellowhouse Canyon Lakes Park, a recreational greenbelt stretching 10 km (6 miles) through the city and covering 580 ha (1450 acres). This park, developed through the wise use of reclaimed wastewater, now provides water-oriented recreational activities in a semi-arid area. Lubbock, like most cities of the Texas High Plains, is using groundwater faster than it can be replaced. With the decline in this natural source, it was vital that the alternative supplies be developed.

26. The use of reclaimed wastewater is not new to Lubbock. For 40 years, the Frank Gray farm, southeast of the city, has been using treated wastewater for crop irrigation. But in the canyon, there is normally little or no natural flow in the stream bed. To ensure that the lakes would be maintained at a constant level, the successive use of reclaimed wastewater was proposed. A study concluded that 44 Mm3 (36 000 ac-ft) of water had been stored beneath the 1600 ha (4000 acres) of irrigated farmland. This treated effluent was superior in quality to other alternatives and less expensive to recover.

27. A series of four dams was constructed to provide lakes within the canyon, control flooding, and allow enough area for adjacent open spaces and parks. The water supply system consists of a series of recovery wells from the irrigation farmland, to the pump station, storage tank-chlorination facilities, and a pipeline 11 km (7 miles) long to the head of the canyon's lakes.

28. Although the lakes use only a portion of the available reclaimed water (15 ML/d [4 mgd]), a nearby power plant uses 27 to 30 ML/d (7 to 8 mgd); the remaining portion of secondary effluent (50 ML/d [13 mgd]) is used for irrigation.

29. San Diego (Calif.) Wild Animal Park. When the 7200 ha (1800-acre) Wild Animal Park opened in 1972 just south of Escondido, some called it a second Kenya. The San Pasqual Valley, where the park is located, has scarce rainfall, warm temperatures, and high veldt (grassland with scattered shrubs and trees) on land that rises from 120 to 560 m (400 to 1400 ft)-all typical of East Africa. Water of high quality and volume was needed to support vegetation, irrigation, and the 3000 animals of 226 different species.

30. Reuse plans were carefully reviewed by veterinarians, the city utilities engineering department, the San Diego Regional Water Quality Control Board, the county and state health departments, and the State Water Resources Department. Treatment now consists of a 190 m3 /d (50 000 gpd) extended

aeration unit. The secondary effluent is chlorinated and released to a holding pond. Other flows provide dilution, but 100% of the park's wastewater effluent is recycled for spray irrigation.

Discussion on Papers 10 and 11

DR A. L. DOWNING, Binnie & Partners, London, UK
In both Papers 10 and 11 we can see that progressive but
pragmatic attitudes have been taken towards reuse of sewage
effluent for recreation. In the Dutch case, which I think is
typical of many situations in Western Europe, the flow of
effluent does not appear to be vital to the attainment of
recreational benefit; the effluent happens to be there and the
problem is how best to control it. In the US cases its use is
vital to the creation of the recreational facility. In both
papers the authors have concentrated on the political,
economic and engineering problems of how best to meet defined
target goals for water quality rather than on reviewing the
rationale (that is to say the dosage–effect relationships) on
which the target standards are based. There is no doubt scope
for debating some of these standards but I shall confine
attention to the means that have been used to meet laid-down
quality goals.

In the case of the Dutch lake it would appear that the
decision to introduce phosphorus (P) removal was taken
probably before the nutrient budget had been fully quantified
and certainly before interactions between nitrogen (N), p and
algae had been modelled. Mr Scheltinga mentions that the p
loading on the lake was 100 t/yr in 1970, that removing 90% of
P from sewage effluent reduces load to about 40 t/yr, and that
the effect has been to reduce chlorophyll content from around
200 mg/m^3 or more to about 100 Mg/m^3. He also notes that
experiments indicated that P was released from sediments at a
significant rate.

This prompts the questions:
- What fraction of the total load was contributed by P
  recycled from the bed and how was this measured?
- Was the reduction in content of algae expected to be
  proportional to the reduction in total load including
  recycle?
- If not, what was expected and how do expectations agree with
  observations?
- Presumably P removal was chosen because it was felt to be
  more economical and reliable than N removal but could the
  author comment on what factors induced the choice?

Reuse of sewage effluent. Thomas Telford Ltd, London, 1984

193

- May I presume that if the same thing had to be done again armed with hindsight the same method would be chosen?
- Having noted earlier in Paper 10 that the engineers were working to meet quality goals expressed in terms of concentration would Mr Scheltinga agree that, as his paper seems to imply, goals for nutrients would generally be more appropriate in terms of load rather than concentration?

It is interesting to note that general information on bacteriological quality is made available in Holland to holiday makers in leaflets. It would be helpful to know something of the terms in which the data are presented. Presumably this is in a style which shows how much below the EC standards the figures are - but one wonders whether people find this reassuring or whether they express any concern that micro-organisms are present at all, especially if the data reveal that the bulk are probably derived from animal manure, rather than sewage effluent that has been disinfected?

Mr Scheltinga reports quite a common experience in that the partitioning of heavy metals between water and suspended matter frequently results in the majority of the metal being adsorbed on solids and being deposited as sludge. Dredged out disposal of this sludge can be difficult. In a UK scheme which could be regarded as indirect reuse the whole flow of a polluted river (the Tame in the Severn-Trent Water Authority area) is passed through a sedimentation lagoon and will eventually proceed to recreational lakes though these have yet to be constructed. Sludge is removed by suction dredger at the rate of about 80 t/day (dry matter). It is pumped (at a solids content around 15%) to a nearby sewage works where it is dried on open beds. The main method of sludge disposal at the works is by incineration but the Tame sludge contains only about 30% volatile matter, and so it is not particularly suitable for incineration. Its high metal content (around 6-7%, mainly iron) renders it unsuitable for agriculture, so it is mixed with municipal refuse and used as landfill.

Even in the most critical period the proportion of sewage effluent appears to be no more than 25%, which is a little below the maximum annual average in rivers abstracted for potable water supply in the UK and well below that in 95% of the situations referred to in Paper 1. Depending on one's point of view this suggests either that the Dutch scheme is fairly conservative in relation to health or that the UK practice for potable supply is rather venturesome.

Paper 11 portrays rather an encouraging picture of situations in the USA where thoughtful planning has evidently resulted in successful schemes permitting recreational use or other benefits to be realized that otherwise would not have been possible at acceptable cost. Several of the schemes mentioned have been operating for 10 years or more and one would imagine that if serious mistakes had been made they would have been revealed by now, except in respect of long-term effects on the health of consumers in cases where aquifers are being recharged.

The authors give an interesting account of their planning
and feasibility assessment procedures and they refer to the
use of computer programs to aid decision making. Could they
enlarge on the nature of these programs? Are these optimizing
models which identify the cheapest sequence when provided with
target quality goals or do they simply calculate the costs of
a sequence specified by the designer? To what extent do they
take account of fluctuations in the flow and composition of
the effluent? I wonder also whether the benefits of improved
amenity are quantified in monetary terms and if so how?

It is interesting to find the standards couched in rather
convenient terms, e.g. twice the potable water standard.
Presumably this is only permissible where it is clear that
substantial dilution (at least x 2) with good quality natural
water would occur before the water reaches the consumer?

Finally, it seems evident that the generous provision of
financial incentives and the mandatory requirement to consider
reuse of sewage effluent in the evolution of new schemes must
be factors favourable to the adoption of reuse; however, the
long list of planning 'hurdles' presented by the author seems
daunting. I wonder where the main opposition to reuse comes
from. I imagine that it is not simply a matter of finding the
scheme with the best benefit:cost ratio. In this context, I
wonder about the role that the ardent conservation groups play
and in particular whether they are usually for or against
reuse. It was of interest to find in a recent planning
enquiry for a reservoir in the UK that such groups were urging
that reuse of sewage should be preferred to devoting land to
formation of a reservoir.

MR H. M. J. SCHELTINGA, Paper 10

I agree that in the Dutch case a rather typical Western
European situation was described - the effluent being
inevitable and recreation being wanted. Indeed, phosphorus
(P) removal from the effluent was based on a rather rough
estimate of several sources. Nor had any model been used to
predict the effect of this choice. The philosophy was that
the large quantity of P in the effluent was at that moment the
only source that could be largely controlled by technical
means. In fact the same approach was followed by improving
the bacteriological quality by chlorination.

Much later, in the later 70s, quantification of P released
from sediments became possible to a certain extent. the
release of P can now be estimated to be 5-10 mg/m^2 per day
during the summer.

The reduction of algae was proportional to the total P
reduction as measured in the water. This was demonstrated
again by the effect of flushing the lake with polderwater,
which is low in P concentration. As a result the average P
concentration in the lake was lowered as was the chlorophyll-a
level and the transparency. The benefits of using animal
manure (in relation to this observation) as a control on
dumping practices has not yet been quantified.

Under the given conditions of the Veluwelake, P indeed is the controlling factor in eutrophication. There is a good relationship between P concentration in the water and the level of chlorophyll. However, nitrogen sources can be attributed mainly to drainage and runoff from cultivated land. They can hardly or not be controlled under present legislation by the water authorities. P removal is already common practice in several parts of Holland to prevent or abate eutrophication. So the choice had already been made! If we had to make the choice today, with our present knowledge, the same decision would have been taken, but at an earlier time in order to prevent the accumulation of P in the sediment.

Goals for a nutrient level in eutrophic waters can, at least in a given situation, be very well expressed in terms of P concentrations, which is fully dependent on retention time and depth of the lake – in this case about 6 months and 2 m.

In the information leaflets for holiday makers the bacteriological data are presented per location as one of three categories: either acceptable to good, or in general unacceptable or uncertain, sometimes unacceptable. These qualifications are comparable with the EC standards, without giving detailed figures.

The Dutch scheme for recreation cannot be compared with the UK practice for potable supply. In preparing drinking water every sophisticated type of treatment can, if necessary, be given in order to comply with quality standards. In recreational schemes for surface water (not in swimming pools) there are only limited possibilities for the control of quality to the benefit of 'consumers'. A fairly conservative approach seems to be justified.

MR J. F. CARUSO and DR R. J. AVENDT, Paper 11
The comments that Dr Downing makes concerning the reuse for amenity projects in the US notes that the projects involve substantial efforts to design facilities to meet various water quality standards rather to the question the rationale for the water quality requirements. We have very limited experience with water reuse for recreational purposes and, therefore, have taken a very conservative approach in establishing effluent limits that may affect public health. As additional data are analysed the effluent quality may be relaxed based on long-term health effects studies. The various computer models that are available to plan and optimize water reuse projects vary in complexity and ease of use. Four computer programs – CHEMTRT, REUSE, LINEAR and PWCST – have been developed to perform the engineering calculations involved in analysing various reuse alternatives. CHEMTRT calculates the effluent water quality produced by various chemical pretreatment processes; REUSE develops design criteria and cost estimates for treatment facilities and effluent distribution system components on a regional basis; LINEAR performs the detailed design, alternative analysis and cost estimation of a particular distribution system; PWCST analyses the present

worth of capital and annual costs associated with a selected
reuse alternative. The use of these models is further
discussed in the Water Reuse Manual of Practice SM-3 published
by the Water Pollution Control Federation, Washington DC
20037. The main opposition to reuse programmes in the US may
very well come from within the regulatory agencies themselves.
The effluent standards are established using a very complex
method of risk assessment that calculates a theoretical
pollutant concentration limit based on lifetime exposure and
an 'acceptable level of risk'. It is very difficult to obtain
the necessary reuse project approval from the general public
when each person at the public hearing believes they are the
one increase in cancer mortality in 10,000,000 that is the
'acceptable level of risk'.

12   Cedar Creek reclamation–groundwater recharge demonstration program

R. J. AVENDT, PhD, PE, Environmental Engineering, Greenhorne & O'Mara, Inc., USA

SYNOPSIS. This paper discusses the background and operation of the Cedar Creek Water Reclamation and Groundwater Recharge Demonstration Program located in Nassau County, New York. This facility is part of a series of local and regional studies funded by Nassau County, the New York Department of Environmental Conservation and the United States Environmental Protection Agency to evaluate wastewater reuse. The results indicate that the potable groundwater supply could be augmented and in some instances enhanced with adequately designed and operated reclamation facilities.

PROJECT PHASES

1. Nassau County located on Long Island, New York, utilizes groundwater as its only source of potable water supply. For this reason, the United States Environmental Protection Agency (U.S. EPA) declared this aquifer a "Sole Source Aquifer". This action requires Agency review of proposed projects that would impact the quantity or quality of the aquifer. Nassau County is currently completing a water reclamation and groundwater recharge demonstration project involving four phases over an approximate twenty year period.

2. The increased urbanization of Nassau County, which is directly east of New York City, has been accompanied by increased groundwater pumpage and the contamination of the upper glacial aquifer. The County is 289 square miles and has a population of approximately 1.6 million. In order to retard the contamination of the upper glacial aquifer and to protect the remaining aquifers (the Magothy and the Lloyd), Nassau County embarked upon an extensive sewering program shortly after World War II. At present, approximately 85% of the County is sewered. The sewage effluent from the treatment facilities, constructed as a part of the sewering program, is discharged to marine surface waters after undergoing secondary treatment.

3. The effect of the sewering program together with the increased groundwater pumpage has been a net decline

Reuse of sewage effluent. Thomas Telford Ltd, London, 1984

199

Table 1. Water Quality Standards For Cedar Creek
Reclamation-Recharge Project

Constituent or Characteristic	Desired Limit mg/l	Constituent or Characteristic	Desired Limit mg/l
ABS	0.5 mg/l	Microbiological	
Aluminum	0.1	Coliform·Org.	4/100 ml. max.
Arsenic	0.05		1/100 ml. avg.
Barium	1.0		
BOD$_5$	2.0		
Boron	1.0		
Cadmium	0.01		
Calcium	<Sat. Con	Pesticides	
CCE	0.2	Aldrin	0.017
Chloride	250	Chlordane	0.003
Chlorine Res. (Free)	1.0	DDT	0.042
Chromium (hexavalent)	0:05	Dieldrin	0.017
Copper	0.2	Endrin	0.001
Cyanide	0.1	Heptachlor	0.018
Fluoride	1.5		
Iron & Manganese Combined	<0.3	Hep. Epoxide	0.018
		Herbicides	0.1
Lead	0.05	Methoxychlor	0.035
Mercury	0.005	Org. PO$_4$+Carbamates	0.1
Nitrogen, Total	3.0	Toxaphene	0.005
O$_2$Consumed	2.0		
Phenols	0.001		
Phosphorus	0.1	Physical	
Selenium	0.01	Color	15 units
Silver	0.05	TON	3 units
Sodium	50%of Cations or 20 mg/l	TDS	500
		Turbidity	0.5 J.T.U.
		Entrained Air	None
Sulfate	250	SS	1.0
TOC	3.0	pH	6.5-8.5 units
Uranyl Ion	5.0		
Zinc	0.3		

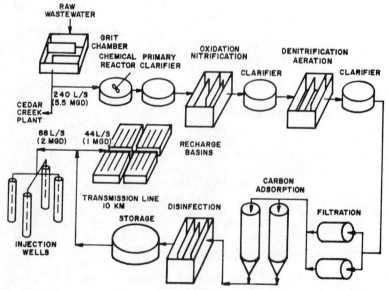

Fig. 1

in the water table which has resulted in a decrease in freshwater streamflow, increases in bay salinity and the local landward movement of sea water into the aquifers. For these reasons, the reuse of water and recharge have been investigated by Nassau County since 1963.

Planning

4. The first phase of the investigations involved numerous planning studies. In 1963, a consulting engineering firm was retained to initiate research into water reclamation and groundwater recharge with emphasis on the construction of a hydraulic barrier to retard salt water intrusion on the south shore of the County.

5. After a series of bench scale studies were completed to provide the necessary design criteria for a pilot facility, a 400 gpm Water Renovation Plant and 500 foot deep injection well was placed into operation in 1968. This facility located at the Bay Park Wastewater Treatment Plant was operated until 1973. The Nassau County Department of Public Works was responsible for the operation of the reclamation facilities while the United States Geological Survey (U.S.G.S) was responsible for the well operation.

6. The evaluation of the operations data for this facility indicated that while water reclamation was feasible; deep well injection was not. This result was due to the fact that recharge took place into the relatively fine Magothy aquifer and was accompanied by a long-term well clogging rate of 3 feet of excessive head build-up per million gallon of injected water. This excessive head build-up indicated that frequent redevelopment of the wells would be necessary to maintain an injection well network. The redevelopment costs would preclude this method of wastewater reuse from being cost-effective.

7. A subsequent consultant's report in 1971, re-evaluated the available means to supplement the Nassau County water supply. The alternatives suggested for further consideration included the importation of water from other neighboring communities and groundwater recharge with reclaimed waste-waters. This planning report also summarized the ongoing investigations of the County Health Department and the U.S.G.S., which indicated that salt water instrusion was not the major concern. Nitrate contamination in the central part of the County was a more serious concern. The report recommended upland recharge of 92 mgd by the year 1990.

8. Based on the results of the above planning reports and pilot investigations, the U.S. EPA, Region II prepared

Table 2. Design Data

```
Incoming Flow Characteristics
   Flow, mgd                                             5.5
   Suspended Solids, mg/l                                240
   Biochemical Oxygen Demand (5-Day), mg/l              220
   TKN, mg/l                                             35
   NH3-N, mg/l                                           22
   Total Phosphorus, mg/l                                16

Lime Treatment System
   Lime Reactor Volume, gal                            4,190
   Detention Time, min                                    1
   Lime Dose, mg/l                                     200 to 350

Flocculation Chamber*
   Number of Chambers                                    2
   Detention Time/Chamber, min                          6.25
   Number of Horizontal Shaft, Slow Speed Mixers         2
   Diameter of Paddle Assembly, ft                       4

Primary Tank*
   Surface Loading Rate, gpd/sq.ft                      830
   Weir Overflow Rate, gal/ft/day                     11,000
   Detention Time, hr                                    2.1

Combined Carbon Oxidation-Nitrification*
   Tank Volume, cu.ft                                180,000
   Detention Time, hr                                    5.9
   MLVSS, mg/l                                         2,700
   F/M                                                   0.15
   Sludge Age, day                                       17
   Oxygen, scfm                                       11,200

Intermediate Clarifier*
   Solid Loading Rate, lb/day/sq.ft                     27.5
   Overflow Rate, gpd/sq.ft                             700

Denitrification-Anoxic Mixed Reactor*
   Volume, cu.ft                                       61,300
   Detention Time, hr                                     2
   MLVSS, mg/l                                          2,700
   Methanol, lb/day                                     3,240

Post Aeration-Stabilization
   Volume, cu.ft                                       30,650
   Detention Time, hr                                     1
   MLVSS, mg/l                                          2,700
   Oxygen, lb/day                                       2,140
   Alum, lb/day                                         3,030

Final Clarifier*
   Solids Loading Rate, lb/day/sq.ft                     36
   Overflow Rate, gpd/sq.ft                             700
   Recycle                                               0.5Q

Filtration
   Number of Units                                    2 (+ 2 standby = 4)
   Unit Size                                          13.3' x 38.7'
   Filter Media                                       Dual Media (Sand,
                                                        Anthracite)
   Filtration Rate, gpm/sq.ft                           3.75
   Backwash Rate, gpm/sq.ft                            15 to 20

Carbon Adsorption System
   Number of Units                                    2 (+2 standby = 4)
   Unit Size, ft                                      13.3' x 38.3'
   Contact Bed Depth, ft                                 8
   Total Carbon, lb                                   482,800
   Unit Loading Rate, gpm/sq.ft                         3.75
   Detention Time, min                                  16 (2 Units in parallel)
   Backwash Rate, gpm/sq.ft                            15 to 20
   Carbon Regeneration Furnace Size          (1)-11.75' Dia. x 5' Hearth
   Carbon Regeneration Furnace Area, sq.ft             450
   Number of Carbon Storage Tanks                        1
   Carbon Storage Capacity, cu.ft                     4,120
```

an Environmental Impact Statement (E.I.S.) The EIS, completed in 1972, recommended that Nassau County study the feasibility of constructing a 5 mgd water reclamation and groundwater recharge facility at its newly completed Cedar Creek Water Pollution Control Plant site.

9. The U.S. EPA then authorized a $95,000 research grant to Nassau County to complete this recommended feasibility study. A consultant was retained to complete the study. The objectives accomplished during the study were discussed in a report entitled "Correlation of Advanced Wastewater Treatment and Groundwater Recharge", which was completed in 1973. This report investigated the feasible methods of recharging water on Long Island. Water quality criteria for a water reclamation-recharge project were then established based on applicable water quality standards, method of recharge, rechargeability of reclaimed water, existing groundwater quality and public health considerations (see Table 1). Knowing the constraints associated with the recharge method and required water quality, numerous wastewater treatment unit processes were critically evaluated. An integrated water reclamation process scheme capable of pro- the required effluent quality was developed based on available current technology. The conceptual design of a reclamation and recharge facility which would serve to demonstrate the performance and reliability of the water reuse system was prepared. The scale and duration of the demonstration program was proposed to be 5.5 mgd for 3 to 5 years.

10. The research report recommended that portions of the Cedar Creek Water Pollution Control Plant be used for the water reclamation process. The report was submitted to the U.S. EPA for approval, and in 1974 was approved. The U.S. EPA authorized the commencement of detailed plans and specifications under the construction grants program. This was a departure from the original plan of conducting the recommended demonstration project under a government sponsored research program. The switch to the construction grants program also deleted several side stream investigations that were proposed for further research. Investigations involving ion exhange, ozonation, fluidized biological reactors and streamflow augmentation were not fundable under the construction grants program. In October, 1974, the County retained a consultant to prepare plans and speci- fications for the water reclamation and groundwater recharge facilities.

Design and Construction

11. The second phase of the project entailed the preparation of detailed plans and specifications for the 5.5 mgd water

Table 3.   Major Equipment Suppliers

Equipment	Manufacturer
Influent and Mixed Liquor Pumps	KSB-Electric Machinery Mfg. Co.
Return Activated Sludge Pumps	Allis-Chalmers
Waste Activated Sludge Pump	Allis-Chalmers
Reclamation Plant Effluent Pump	Allis-Chalmers
Carbon Adsorption Tank Effluent Pump	Allis-Chalmers
Carbon Slurry Pump	Marlow Pumps
Ferric Chloride, Polymer & Alum Feed Pump	The Madden Corp.
Lime Slurry Pump	Dorr-Oliver, Inc.
Sampling Pump	Robbins & Myers
Granular Polymer Mixing System	Acrison, Inc.
Lime Slaking System	Wallace & Tiernan Div. Pennwalt Corp.
Lime Conveying System	Rage Engineering, Inc.
Package Pumping Stations	Lyco-ZF
Methanol Pumping Station	Lyco-ZF
Fiberglass Tanks	Owens-Corning
Steel Tanks	Eastern Tank Fabricators, Inc.
Gravity Filters	Leopold/Sybron Corp.
Carbon Media	Westvaco
Carbon Regeneration System	Envirotech
Flocculators	E&I Corp.
Chlorination Mixers	E&I Corp.
Instrumentation	Robert Shaw
Butterfly Valves, Operators & Appurtenances	Centerline
Slide Gates	Tyler Pipe Ind.
Major Gate Valves, Operators & Appurtenances	Dresser Ind., Fairbanks, Kennedy
Major Check Valves	Clow Corp.
Laboratory Equipment	Hamilton Industries
Hoist, Trolley & Monorail System	Elbert Lively & Co., Inc.

reclamation-recharge facility and its construction at the existing 45.0 mgd Cedar Creek Water Pollution Plant. The project can be divided into three main components for discussion: water reclamation facility, transmission main and recharge facility.

12. Water Reclamation Facility. The design approach taken for the 5.5 mgd water reclamation facility incorporated the best available demonstrated technology with maximum utilization of the existing plant facilities to insure consistent production of the high quaity effluent required for recharge. Sufficient operational flexibility was provided to enable optimization of operation and control parameters. The facility has continuous automated monitoring and alarm systems, an emergency power supply, short term retention of the reclaimed water, ocean outfall disposal as an alternative discharge, a full complement of standby equipment, and multiple units.

13. The water reclamation facility is shown in overall schematic in Figure 1. Design data and operating parameters are listed in Table 2. The water reclamation processes used include chemically aided primary sedimentation, two or three state biological treatment, filtration, carbon adsorption, chlorination and storage. Major equipment suppliers are listed in Table 3.

14. Chemically Aided Primary Treatment. The primary treatment system employs addition of lime followed by primary sedimentation at pH 9.5. This increases removal of phosphorus, BOD_5, suspended solids and heavy metals, thereby optimizing and protecting the biological nitrification system.

15. The influent flow to the water reclamation plant is screened and degritted wastewater withdrawn under controlled conditions from the existing plant primary influent flow. The wastewater is conveyed to chemical treatment facilities. Contained in the chemical treatment facilities are a rapid mix tank for dispersal of the lime, automatic lime slaking and feeding equipment, a slurry tank and the required control and monitoring devices. The lime mixed effluent from the rapid mix tank is discharged by gravity to two flocculation chambers. In each of the chambers a horizontal shaft slow speed mixer is installed for flocculation. The flocculated effluent then flows to one of two primary tanks that are used in the reclamation process.

16. Two Stage Biological Treatment System. The secondary treatment system has been designed to promote biological nitrogen removal in conjunction with oxidation of the remaining

carbonaceous material. The design allows for the operation of the secondary system as a two stage suspended growth system, with combined carbon oxidation-nitrification, intermediate clarification, denitrification, post aeration-stabilization and final clarification. The settled wastewater from the primary tank flows by gravity to existing aeration tanks. A portion of the existing aeration tanks provides a detention time of 5.9 hours based on the design flow rate. From the combined carbon oxidation-nitrification tank the mixed liquor enters one existing 100 foot diameter final clarifier. Return activated sludge is returned to the carbon oxidation-nitrification tank, utilizing the existing return sludge well, pumps, meters and lines. The nitrified effluent from the intermediate clarifier passes to the denitrification tank which provides detention time of two hours. Methanol is added to the tank influent flow at a controlled rate as a supplemental organic carbon source in the denitrification process.

17. The denitrified liquor flows to a post aeration-stabilization tank. The purpose of the post aeration-stabilization tank is to remove any supersaturated nitrogen gas, to avoid rising sludge problems in the final clarifier, to provide a mixing zone for the addition of alum and/or polymer for further phosphorus removal, and to provide an additional aeration period for the removal of excess methanol. The aerated liquor is conveyed to an existing clarifier for final clarification. Operational flexibility is provided to allow conversion from a two stage to a three stage biological treatment system as a fail-safe mechanism to assure the production of a high quality effluent. A three stage intermediate system would consist of a high rate activated sludge unit, separate nitrification, followed by denitrification.

18. Filtration and Carbon Adsorption. The filtration and carbon adsorption system design is based on the use of multiple individual units, the size of which can be incorporated directly into a future fullscale Nassau County recharge system. The clarified effluent from the secondary system is pumped under controlled conditions to automatically backwashed, dual-media filters. Two gravity type filters are provided with two additional filters on standby. Each filter provides a surface area of 515 square feet and operates at under 4 gpm/ft^2.

19. The carbon adsorption facilities consist of two contractor units in parallel, with two standby units provided. Each unit provides 515 square feet of surface area. The adsorption units are operated as downflow packed bed units, 8 feet deep. Contact time is approximately 16 minutes.

20. Chlorination and Storage. Two unused aeration tanks providing 9 hours detention time were covered and serve as the chlorine contact chamber and storage tank. The storage tank also serves as a pump suction chamber for high head pumps to deliver treated water to the recharge site and provides the necessary equalization to permit continuous delivery of 4.0 mgd to the injection wells, despite fluctuations in plant output due to backwashing of filters and carbon adsorbors.

21. Transmission Main System. The demonstration project involves the use of 24 inch diameter concrete pipe transmission main to convey reclaimed water the Meadowbrook recharge site. The routing parallels the Wantagh State Parkway from the Cedar Creek treatment plant to the recharge site. The length of transmission main is approximately 6.25 miles.

22. Recharge Facilities. The Meadowbrook recharge site is a triangular-shaped piece of County-owned land located in the geographic center of the County. It includes approximately 60 acres of land which is occupied in part by the Nassau County Correctional Center and the former Meadowbrook Sewage Treatment Plant.

23. Both shallow well and basin recharge studies are to be conducted at the Meadowbrook site. The proposed program will recharge 4 mgd, with approximately 2.0 mgd recharged through basins and the remaining through wells. Five wells and eleven basins are available. The basins, each 50' x 100' will be used to determine the management practices that are most effective for optimizing recharge. Evaluation of the clogging phenomena associated with the application of reclaimed wastewater and the potential for the unsaturated zone to improve water quality will be performed. Five wells, 100' deep with 30' screens are installed at the Meadowbrook site to recharge reclaimed wastewater. Four of the wells may be in operation any specified time, with one on standby. Each well would recharge 0.5 mgd.

24. The Meadowbrook recharge site is surrounded by a monitoring network of 46 observation wells. They are installed near the center of both the injection well and basin recharge sites. In addition to the observation well network, an onsite reservoir is automatically monitored.

25. The plans and specifications were completed in early 1976. On March 24, 1976, a Public Hearing was held to discuss the result of the Environmental Assessment Statement (EAS) for the project. As there were no objections voiced at this hearing, the contract documents, including plans and specifications, were submitted to the State and Federal agencies for approval. In June, 1976, Nassau County was

Table 4. SPDES Final Effluent Discharge Limitations

Effluent Parameter	Discharge Limitations (Daily Max.)	Effluent Parameter	Discharge Limitations (Daily Max.)
Metals	mg/l	**Base/Neutral/Acid Compounds**	ug/l
Antimony	0.05	Bis (chloroethyl) ether	0.09
Arsenic	0.025	(3) 1, 4 Dichlorobenzene	10.0
Barium	1.00	(3) 1, 2 Dichlorobenzene	10.0
Beryllium	0.20	Nitrobenzene	30.0
Cadmium	0.01	Bis (2-chloroethoxy) methane	2.0
Chromium - total	0.5	(6) Naphthalene	30.0
Colbalt	1.00	(6) 2-Chloronaphthalene	10.0
(1) Iron	0.30	(6) Acenaphthylene	10.0
Lead	0.025	(7) Diethylphthalate	50.0
(1) Manganese	0.30	2, 6 Dinitrotoluene	50.0
Mercury	0.002	(6) Fluorene	10.0
Nickel	2.0	(7) Diethylphthalate	50.0
Selenium	0.01	Azobenzene	1.0
Silver	0.05	N-Nitrosodiphenylamine	14.0
Thallium	0.02	(6) Phenanthrene	10.0
Zinc	0.30	(6) Anthracene	10.0
Copper	0.20	(7) Di-n-butylphthalate	50.0
		(6) Fluoranthene	0.2
Volatile Organics	ug/l	(6) Pyrene	0.2
		(7) Butylbenzylphthalate	50.0
(2) Dichlorofluoromethane	50.0	(7) Bis (2-ethylhexyl) phthalate	75.0
Chloroethene	1.0	(7) Di-n-octylphthalate	50.0
(2) Dichloromethane	50.0	(3) 1, 3 Dichlorobenzene	10.0
(2) 1, 2 Dichloroethene	1.0	N-Nitrosodiphenylamine	14.0
(2) 1, 1 Dichloroethane	50.0	(8) Phenol	1.0
(2) Trichloromethane	10.0	(8) 2, 4 Dimethylphenol	1.0
(2) 1, 1, 1 Trichloroethane	50.0	(8) 4-Chloro, 2-methylphenol	1.0
1, 2 Dichloroethane	2.0	(8) 2, 4, 6 Trichlorophenol	2.0
(2) Tetrachloromethane	0.3	(8) 2-Nitrophenol	1.0
Benzene	(9)	(8) 2, 4 Dichlorophenol	1.0
Trichloroethene	5.0	(6) Acenaphthene	10.0
(2) Bromodichloromethane	50.0		
(5) Methylbenzene	50.0	**Misc. Chemical**	mg/l
(2) Dibromochloromethane	50.0		
Tetrachloroethene	10.0	Ammonia	2.0
(3) Chlorobenzene	10.0	Chloride	250.0
(5) Ethylbenzene	50.0	Fluoride	1.5
(2) Tribromomethane	50.0	Nitrates + Nitrites	10.0
		Cyanide	0.2
Chlorinated Compounds	ug/l	Chlorine Residual - Free	5.0
		Sodium	170.0
Alpha BHC	0.2	Oil and Grease	15.0
Beta/gamma BHC	0.2	pH	(10)
Delta BHC	0.2		
Heptachlor	(9)	**Physical**	
Aldrin	(9)		
Dieldrin	(9)	Turbidity	5 NTU
Heptachlorepoxide	(9)	Color	15 Units
(4) P, P' - DDE	0.1	TDS	1,000 mg/l
Endrin	(9)	Flow	4.0 MGD
Beta-endosulfan	5.0		
(4) P, P' - DDD	0.1	**Microbiological**	
Endrin aldehyde	(9)		
Endosulfan sulfate	5.0	Virus	(11)
(4) P, P' - DDT	(9)	Coliform	1/100 ml
(4) P, P' - DDT	(9)		
Mirex	0.1		
Alpha Chlorodane	0.03		

NOTES:

(1) to (8) Total aggregate concentration of these constituents shall not exceed:
(1) 0.50 mg/l, (2) 50.0 ug/l, (3) 10.0 ug/l, (4) 0.1 ug/l, (5) 50 ug/l,
(6) 30 ug/l, (7) 75 ug/l, nor (8) 1.0 ug/l.
(9) Not detectable by tests or analytical determination references in 6 NYCRR 703.4.
(10) The pH shall not be less than 6.5 standard units nor greater than 8.5 standard units.
(11) No viral standard has been set, although analyses must be conducted on a weekly basis.

notified that it was awarded a U.S. EPA construction grant. This grant award was for both the design and construction portions of the demonstration program. Costs under this grant were shared as follows: U.S. EPA-75%; New York Department of Environmental Conservation (NYDEC) -12 1/2% and Nassau County 12 1/2% considering a total project cost of $25,795,739. In November, 1976, the U.S. EPA and NYDEC approved the plans and specifications for the project.

Operations

26. Construction of the water reclamation plant was scheduled for two years. The transmission line and recharge facilities were scheduled for eighteen months construction periods. Normal construction delays and acceptance of the instrumentation equipment delayed the start-up of the water reclamation plant until April, 1980.

27. During the construction, it was decided to finalize the necessary permits for the demonstration program. The original intent of the project was research oriented, however, the construction grants program funding required that a permanent facility be constructed. A New York DEC discharge permit was applied for in early 1978. Initial contact with U.S. EPA, DEC and local health department officials indicated that the necessary State Pollutant Discharge Elimination System (SPDES) permit would not be an early task. The chemical and physical parameters to be listed in the permit were not the cause of concern. The New York State Drinking Water Standards, Part 72 or better were used for the original proposed standards (Table 1). These standards were utilized to select and design the various unit process for the reclamation plant.

28. The problems with finalizing the permit were caused, in part, by the time lapse between the preliminary design of the reclamation plant and operation. Improvements in analytical chemistry and the health sciences had resulted in identifying numerous organic compounds potentially hazardous to health. Drinking water standards and discharge permits were being expanded to address these contaminants. It was agreed to perform a series of organic analyses on the reclamation plant influent and effluent and to list those organics identified in the SPDES permit. In addition, all unidentified peaks which were found in subsequent analyses were to be identified and reported. Table 4 is a tabulation of the contaminant levels contained in the SPDES permit which was subsequently issued March, 1982. Prior to receiving the discharge permit, the reclamation plant was operated for fifteen months with discarge through the Cedar Creek ocean outfall.

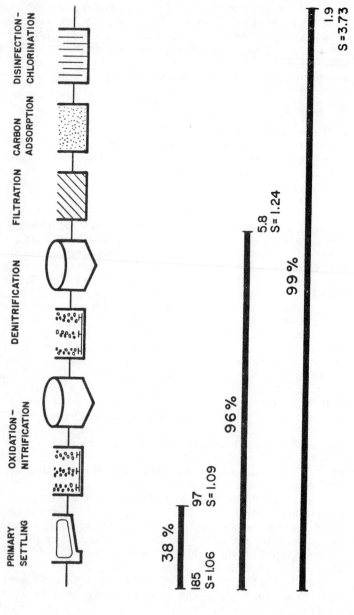

Fig. 2. BOD$_5$ removal

29. The operation and maintenance of the water reclamation and recharge facilities were performed by Nassau County employees. Due to the highly technical nature of the demonstration project, the County retained consultants to optimize the reclamation plant operations, perform virological and microbiological work, provide trace organic analysis and laboratory quality assurance. The U.S.G.S. was responsible for operation of the recharge facility under a cooperative study agreement with the County.

30. The water reclamation plant was placed into operation in April, 1980. Reclaimed water was recharged into the groundwater starting March, 1982. During the operation of the demonstration program, the wastewater flows tributary to the Cedar Creek plant continued to increase as a result of sewer extensions. The water reclamation facilities utilized major portions of the secondary plant for chemically aided primary sedimentation, combined carbon oxidation-nitrification, denitrification, chlorination and storage. Additional secondary sedimentation tanks were constructed in 1979 in an effort to reduce the overload on the remaining secondary treatment plant. The operation of the biological systems was maintained as a two sludge system in an effort to reduce the aeration demands on a centralized blower system.

31. The design capacity of the Cedar Creek secondary wastewater treatment plant was exceeded in early 1979. The continued operation of the water reclamation plant reduced the secondary treatment plant performance and reliability. In 1983 the secondary treatment plant was unable to maintain the required treatment efficiency and consideration was given to suspend the operation of the demonstration program. Secondary treatment levels continued to deteriorate throughout 1983. The overload conditions on the secondary treatment units resulted in periodic ordor problems. In order not to further deteriorate the secondary treatment plant performance, the reclamation-recharge operations were formally terminated in February, 1984. Conversion of the reclamation plant primary and biological systems back to conventional secondary facilities was completed by mid-1984.

Performance Evaluation

32. Currently Nassau County and the U.S.G.S. are conducting comprehensive evaluation programs to determine the performance and reliability of the reclamation and recharge studies. Characterization of the raw wastewater and reclaimed water prior to recharge indicated that priority pollutants are not present, or at levels not warranting concern. The overall performance and reliability of the water reclamation

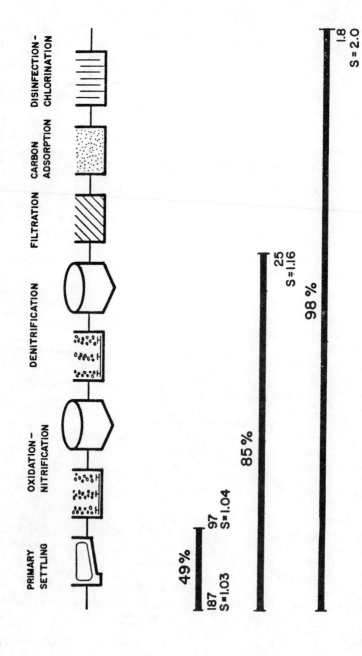

Fig. 3. Suspended solids removal

plant is currently being evaluated. The results of the first fifteen months of operation have been thoroughly evaluated in order to optimize the reclamation plant operation and investigate treatment process stability.

33. An investigation of the available methods to evaluate performance data considered data plots vs. time, relative frequency plots, normal distribution plots was selected to analyze the relationships between design, operation and peformance. This method allowed evaluation of performance on a probability basis with the geometric mean value approximated at the 50% probability (or median) level. When the data is in statistical control the geometric mean is a fractional value approaching the arithmetic mean.

34. An investigation of the available methods to evaluate the reliability of the unit processes considered plotting the standard deviation of performance data, 95% confidence intervals, R Charts, the relative standard deviation, coefficients of variation, and the Spread Factor. The Spread Factor, S, is the antilog of the standard deviation of the logarithms of the original data on the lognormal frequency distribution plot. This measure was selected to analyze the relationships between design, operation and the reliability of performance. Since the Spread Factor represents the slope of the log normal probability plot, the performance and reliability can be determined from the same figures.

35. A summary of the performance/reliability analysis is presented below for several conventional pollutant removals through the reclamation plant.

36. **BOD$_5$:** The removal of BOD$_5$ across the plant averages 98%. Analysis of the spread of data indicates that the performance of the chemically aided primary sedimentation tank is quite stable (38% removal), as indicated by the low S values of 1.06 and 1.09, respectively. BOD$_5$ removal across the plant through biological treatment exceeds 96% comparing geometric means and results in an increased spread of effluent values, S = 1.24. Carbon adsorper effluent values have a geometric mean of less than 2 mg/l with S= 3.73. It appears that the performance of the reclamation plant to remove BOD$_5$ is greater than anticipated. However, the higher the performance the more relative spread in the effluent values (Figure 2).

37. **Suspended Solids:** Suspended Solids (Figure 3) removal in the reclamation plant exceeds 98%. Again the chemically aided primary sedimentation tank is very stable with 49% removal based on geometric means and equal S factors.

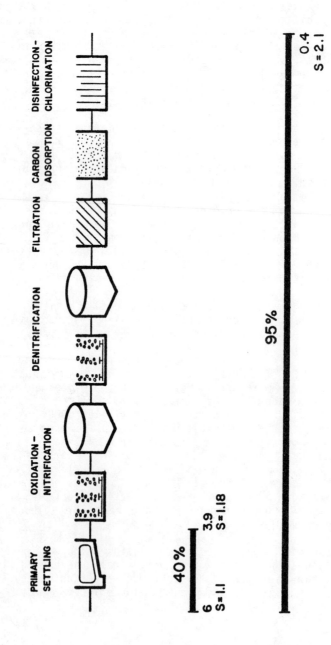

Fig. 4. Phosphorous removal

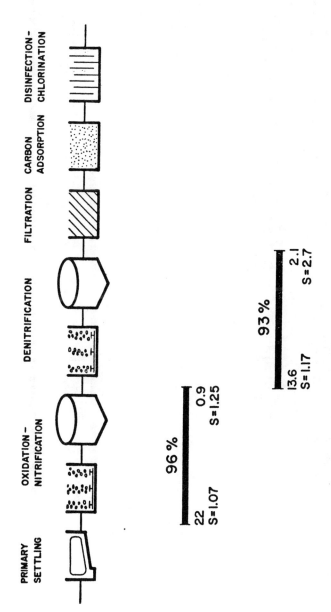

Fig. 5. Nitrogen transformation

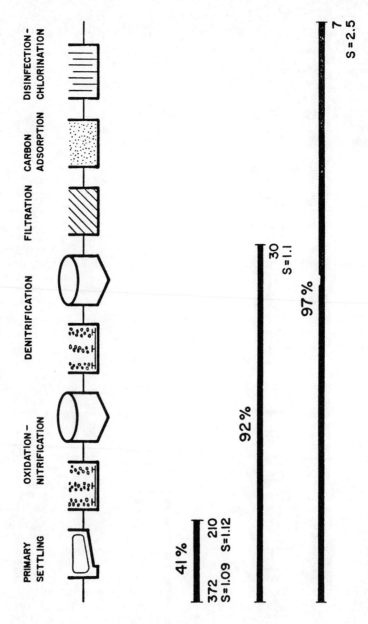

Fig. 6. COD removal

The removals across the biological units and through filtration and adsorption are stable with S values equal to or less than 2.0.

38. **Phosphorus:** Phosphorus removal is accomplished through chemical precipitation/sedimentation and filtration. A 40% reduction is achieved on a stable basis (S=1.18) across the chemically aided primary tanks with the lime addition to pH 9.2. Overall reduction exceeds 95% (geometric means) and is relatively stable (S=2.1). See Figure 4.

39. **Nitrogen Transformation:** Oxidation of ammonia nitrogen to nitrate is accomplished in the first stage biological system. Essentially complete nitrification occurs (see Figure 5). The overall rate is 96% (geometric) and is quite stable (S=1.25). It appears that the system is very conservatively designed and no appreciable effect of temperature is noticed. Denitrification in the second stage biological system utilized methanol as the supplemental organic carbon source. The process removes 93% of the nitrate nitrogen producing an effluent averaging 2.1 mg/l (geometric). The denitrification system (S=2.7) is less stable than nitrification. Methanol dosage has been kept at approximately 3.8 to 1 on weight basis.

40. **Chemical Oxygen Demand:** Overall reduction of COD is 97% (geometric), as shown in Figure 6. Removals across the primary tanks is quite stable (S=1.12) and averages 41% (geometric). Through the biological secondary, the reduction is just as stable (S=1.1) and averages 92% (geometric). The final effluent spread is relatively unstable (S=2.5) with a geometric mean value of 7 mg/l.

SUMMARY AND CONCLUSIONS

41. Evaluation of the operations data is only one measure that is being evaluated to determine the overall success of the Cedar Creek Water Reclamation and Groundwater Recharge Demonstration Program. Currently the U.S.G.S. is studying the recharge water plume as it moves away from the recharge site. Additional observation wells were provided to monitor the short term transport and transformation of contaminants contained in the recharge water. The unique below grade laboratory stations constructed in two of the recharge basins were also used to evaluate contaminant transformation through the unsaturated zone beneath the basins.

42. The fail-safe nature of the reclamation-recharge demonstration program is also being evaluated. The further implementation of reuse will depend on the ability to maintain regulatory compliance, reduce costs, provide public confidence

and minimize adverse environmental impacts. Each of the design and operational parameters critical to the performance and reliability of the water reclamation and recharge operation are being identified. A risk analysis is being performed to evaluate the back-up facilities and maintenance requirements for continuous vs. intermittent operation. Spare part inventories, preventative maintenance and usefull life will be reviewed as they impact fail-safe operation Capital and Operational Cost models are being developed to evaluate the expense of future water reuse programs. Both fixed and variable costs will be indexed to allow cost projections.

44. A comprehensive summary of the demonstration program from planning to phase out is currently being prepared. This program summary will present the goals and objectives of the demonstration program in terms relative to regional water supply planning reports and groundwater resource management. The costs and environmental consequences of full-scale municipal water reuse will be discussed in terms of immediate and long term impacts. The demonstrated performance and reliability of the water reclamation and recharge. The results of the virological and priority pollutant studies will be correlated with the reclamation plant performance comparisons will be made with the observed removals at other water reclamation facilities.

14 Groundwater recharge of sewage effluents in the UK

H. A. MONTGOMERY, PhD, CChem, FRSC, MIWPC, Consultant, M. J. BEARD, LRSC, MIWES, Southern Water Authority and K. M. BAXTER, BSc, MSc, MIGeol, FGS, WRC Environment, Marlow, Buckinghamshire, UK

SYNOPSIS. Groundwater quality investigations at nine effluent recharge sites in the U.K. are reviewed. Good removal was observed of organic matter, ammonia, bacteria, and viruses, and also of total nitrogen at Chalk and gravel sites receiving crude or settled sewage. Little nitrogen was removed at Triassic sandstone sites and at Chalk sites receiving biologically treated effluent. More information would be desirable on trace organic compounds, and on certain operational and engineering aspects of effluent recharge.

INTRODUCTION

1. About 300 Ml/d of sewage effluent are recharged to the ground in the U.K., about one half of it to the Chalk and much of the remainder to the Triassic sandstones. Until recently this practice was regarded merely as a convenient way of disposing of sewage, but this paper seeks to show that, in the right circumstances, recharge can be both an effective means of purifying sewage and a method of conserving water resources.

2. Concern about possible adverse effects of effluent recharge upon groundwater quality led the Department of the Environment to commission research by the Water Research Centre in the middle 1970's, and four regional Water Authorities (SWA, STWA, WNWDA and AWA) started their own research at about the same time. Some of this work was supported by the EEC. Contamination of parts of the aquifer exploited for potable water could make it necessary to abandon sources or introduce expensive treatment, and pollution of hitherto unexploited groundwater reserves could interfere with their future development for water supply. This paper reviews the published results of studies of the effects of effluent recharge upon groundwater quality in the U.K. at the sites shown in Fig. 1; the infiltration rates quoted on Fig. 1 are nominal, and actual hydraulic loadings will have been higher depending on the fraction of the total recharge area in use. A comprehensive review covering effluent recharge world-wide, including most of the sites discussed in this paper, has recently been published by Baxter and Clark (ref. 1).

Reuse of sewage effluent. Thomas Telford Ltd, London, 1984

219

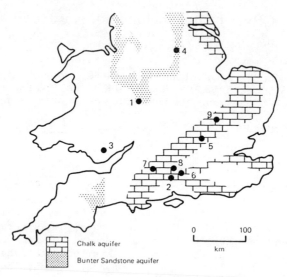

Site No.	Area (m² x 10³)	Nominal infiltration rate (see text) (m/d)	Effluent type
1 Whittington	1020	0.012	
2 Winchester	430	0.026	Primary
3 Cilfynydd	14	0.175	
4 Worksop	1500	0.0045	
5 Caddington	4	0.22	
6 Alresford	23	0.031	Secondary
7 Ludgershall	50	0.41	
8 Whitchurch	15	0.08	Comminuted
9 Royston	137	0.012	Primary + 27% Industrial

Fig. 1. Location of effluent recharge sites investigated

CHALK SITES

3.   Much of Southern and Eastern England relies on the Chalk aquifer for supplies of water.   Numerous village sewage works discharge to the Chalk but the largest site is probably that at Winchester (approximately 11 Ml/d).   The Chalk is a soft lime-stone with many micro-fissures and also with large fissures which are prevalent in particular areas, especially near the Chalk/Tertiary boundary.

4.   Ludgershall (ref. 2).   At this works in East Wiltshire about 0.7 Ml/d of biologically treated sewage effluent is released from orifices in a distribution channel across a shall-ow porous topsoil directly overlying the Chalk.   The hydraulic performance is excellent, a loading of the order of 10^4 mm/d being applied.   Chemically the changes occurring in the unsat-

urated zone of 10-20 m comprise removal of most of the residual
BOD, phosphate and ammonium, and slight, localised denitrification. Coliform bacteria are mostly removed in the unsaturated zone and are almost completely absent after a few hundred
m travel. No other changes were observed except those attributable to simple dilution by the native groundwater.

    5. Alresford (ref. 3). This is a village sewage works near
the headwaters of the River Itchen in Hampshire. About 0.6
Ml/d of biologically treated effluent are discharged to an
unsaturated zone of approximately 20 m through a French drain
network embedded in gravel. The investigation was primarily
concerned with the removal of bacteria. As at Ludgershall,
excellent removal of coliform bacteria was observed, by better
than 4 orders of magnitude a few hundred m from the recharge
site, with the unsaturated zone being more effective than the
saturated zone per unit distance traversed. Chemical changes
were slight, apart from phosphate removal which was virtually
complete. The gravel pack surrounding the French drains gradually becomes blocked at this site and major maintenance is
required at intervals of the order of 10 years.

    6. Winchester (ref. 2). Sewage from the city is pumped to
near the top of St. Catherine's Hill where it is treated by
sedimentation only and then descends the hillside, soaking away
via a series of trenches and lagoons. The effluent mixes with
the native groundwater and enters the River Itchen by diffuse
upward flow through alluvial deposits along several km length
of the river valley. Removal of BOD in the unsaturated zone
(which ranges from 4 m to 36 m in depth) is similar to that
occurring in conventional treatment by biological filtration
but removal of inorganic nitrogen is mostly more than 50% and
sometimes as much as 90% suggesting that the descending effluent
traverses alternate aerobic and anaerobic micro-zones where
nitrification is followed by denitrification. Further purifica
-tion takes place in the saturated zone and careful measurements
have failed to show any effect of the effluent on nitrogen or
phosphate levels in the River Itchen. Coliform bacteria and
viruses are well removed. Dichlorobenzene is the principal
identifiable organic compound, definitely derived from the
sewage, to escape removal in the Chalk.

    7. At the end of 1982, it became necessary to reconsider the
future of the Winchester works when it became known that the
proposed line of a motorway extension would bisect the works
and remove one third of the recharge area including the lagoons.
Experiments have been put in hand, to determine the soakage
capacity of adjacent land, designed to test both ditch and
surface irrigation. Subject to a successful outcome to this
investigation, it is probable that the principle of primary
effluent recharge will be retained, since the advantages over
the option of secondary treatment and river discharge are seen
as :-

        (i) Extensive purification which offers total protection
             from pollution risk to the water supply intake 3 km
             below Winchester.

(ii) No capital or revenue expenditure is incurred in the
respective construction and operation of secondary
treatment facilities.

8. Whitchurch (ref. 2).   Crude sewage (approximately 1.3
Ml/d) was discharged into trenches at this site near the River
Test until 1981 when a biological treatment works discharging
to a French drain system was built.   Anomalies in the hydro-
geology made the results more difficult to interpret than at
the sites discussed previously.   Nevertheless, there was a
clear pattern, as at Winchester, of good removal of BOD, TOC,
and bacteria during the crude sewage discharge regime.   Methane
gas was detected in the unsaturated zone, proving the existence
of anaerobic biological activity, and this probably accounted
for the removal of 26-43% of the total dissolved nitrogen
applied prior to 1981 (ref. 4).   Dichlorobenzene was again the
only trace organic compound, unambiguously of sewage origin, to
persist, and 1 μg/l was present in the groundwater 400 m from
the recharge area.

9. Since the changeover to biological treatment late in 1981
there has been a gradual improvement in groundwater quality at
boreholes close to the soakaway area with BOD values falling and
ammonia being replaced by nitrate.   There has been little
change in water quality at the more remote boreholes.   Concentrations of total nitrogen are almost unchanged as the removal
previously caused by denitrification in the ground is now
effected by sedimentation in the treatment plant.

10. Caddington (ref. 5).   About 1 Ml/d of biologically
treated sewage effluent is discharged through a French drain
system at this site in the Chilterns near Luton.   The effluent
is chlorinated as a precaution because there is a borehole for
public water supply 2.5 km down-gradient.   Chemically the
results were very similar to those described above for
Ludgershall.   Despite the chlorination, the overall reduction
in bacterial numbers was no better than that observed at the
Hampshire and Wiltshire sites and traces of chlorinated hydrocarbons were formed in the groundwater.

11. Royston (ref. 6).   Until 1979 when a treatment works
discharging to surface waters was provided, this discharge in
South Cambridgeshire consisted of 1.8 Ml/d of partially settled
sewage of high chloride content which was recharged through
lagoons.   The results of groundwater quality investigations
were very similar to those already described for Winchester and
Whitchurch.

TRIASSIC SANDSTONE SITES

12. The Triassic sandstone is the second major aquifer in the
U.K. and within the Severn-Trent Water Authority area alone
14.5% of drinking water is abstracted from this aquifer.   It is
more variable than the Chalk ranging from unconsolidated sands
to cemented, fractured sandstones.

13. Whittington (refs. 1, 7).   This site is in the Stour
Valley in the West Midlands and has received an average DWF of
some 12 Ml/d of mixed primary and industrial effluent for the

last 90 years.   Surface irrigation occurs from some 120 to 150
chambers located across the site, connected by four buried
carrier mains.   Sluice gates in these chambers enable the area
of effluent recharge to be varied to suit the needs of the
farmer.   The unsaturated zone is 25-30 m thick and the fractured sandstones have a transmissivity of between 500 and 1000
m^2/d.   The treatment capacity of the Triassic sandstone would
appear to differ from that of the Chalk in that the groundwater
is highly contaminated to about 130 m beneath the site, although
the average hydraulic loading (0.012 m/d) is less than at most
of the Chalk sites where such extensive groundwater contamination is not observed.   There is negligible removal of total
nitrogen over the saturated zone and similar nitrate contamination of Triassic sandstone groundwater has been recorded at a
second effluent recharge site just to the north of Whittington
(ref. 7).   The cause of this reduced nitrogen removal is not
known, but may be related to the higher storativity of the sandstone allowing greater oxygen penetration thereby preventing
the formation of the anoxic conditions necessary for denitrification.

14.   Heavy metals are largely removed during recharge but
levels of lead in the interstitial water of the saturated zone
suggest that lead may be particularly mobile.   The removal of
enteric bacteria and viruses observed beneath the recharge area
is 100% for both bacteria and viruses, unlike the Chalk sites
where such good removal is only found some 300-500 m down-
gradient.   Whole rock analysis at this site has shown that
effluent recharge has leached calcite from the aquifer itself.

15.   Worksop (ref. 8).   About 7 Ml/d of settled sewage were
applied to permanent pasture at this site in Nottinghamshire,
by surface irrigation, for many years until a treatment works
discharging to the adjacent river was opened in 1976.   Concern
about rising nitrate levels at local boreholes led to a detailed
investigation which showed that the groundwater was contaminated
with nitrate down-gradient of the recharge site, upto 50 mg/l
(as N) being observed at the borehole closest to the site.   The
implication was that most of the nitrogen content of the sewage
entered the groundwater as nitrate and that nitrogen removal
was negligible, as at Whittington but unlike the Chalk sites
receiving settled or crude sewage.

ALLUVIAL GRAVEL SITE

16.   Twenty-two per cent of effluent recharged in the U.K.
enters minor aquifers.   Although such aquifers are not of
strategic importance for water resource purposes they may nevertheless be locally important.

17.   Cilfynydd (refs. 1, 9).   This site in South Wales has no
water resource significance and the investigation by WRC and
the then WNWDA was designed to determine effects of effluent
recharge upon groundwater quality in an alluvial gravel aquifer
(actually containing sands, silts and clays in addition to
gravel).   About 2.25 Ml/d of primary effluent, of mixed
domestic and industrial origin, has been allowed to flow over a

series of fields on the river terraces of the River Taff for the last 90 years. Infiltration occurs through three main ditches, which are in continuous use, each leading to a series of secondary ditches. The locations of these secondary ditches vary across the recharge area, a new channel being cut when an existing one becomes blocked.

18. Considerable treatment of the recharged effluent is achieved thereby reducing the effect of the effluent on the quality of the River Taff. Much of the effectiveness of this system would seem to be due to the changes made in the recharge area which allow the unsaturated zone to 'rest' and regenerate its treatment capacity. Nearly 60% of the inorganic nitrogen and 30-50% of the organic carbon are removed. Phosphate and heavy metals, other than iron and manganese, are removed within the first 0.1 m of the soil zone. Concentrations of iron and manganese, however, seem to increase as a result of being leached from the soil/alluvium.

DISCUSSION

19. The studies reviewed in this paper have shown that the aquifers receiving recharge, and particularly their unsaturated zones, act in an analogous way to a biological sewage treatment plant in removing organic matter and in nitrification. In the Chalk and the alluvial gravel, but not the Triassic sandstone, there is the bonus of partial nitrogen removal at sites where the effluent does not receive prior biological treatment. Removal of bacteria and viruses is virtually complete in the unsaturated zone of the sandstone and within 500 m of the recharge sites in the Chalk. The results for Chalk and alluvial gravel are similar to those found for generally unconsolidated aquifers in the U.S.A. (e.g. refs. 10, 11). There would appear to be no advantages in providing biological treatment before recharge, except possibly for aesthetic reasons. The water quality investigations reviewed were not comprehensive in respect of trace organics and of virus removal and more information on these points would be desirable.

20. Water resource implications of sewage effluent recharge will generally be minor except that the recharge will obviously contribute to the base flow of the nearest river down-gradient, with fluctuations in flow and composition being smoothed out compared with the discharge of the same effluent directly to the river. The decision to retain effluent recharge at Winchester in preference to a discharge to the river (probably needing tertiary treatment), or a trunk sewer leading out of the catchment, will not only save some or all of the cost of such schemes but, compared with the trunk sewer scheme, will conserve a water resource of 11 Ml/d valued in the region of £1 M.

21. Recharge of coastal aquifers to provide a barrier between salt and fresh water, and reduce losses of fresh water of good quality to the sea, has often been suggested. The studies reviewed in this paper suggest that, with good planning and management , there need be no objections on water quality

grounds to such a system.

22. The failure to remove nitrogen by the Triassic sandstone resulted in unacceptable increases in the aquifer near Worksop but it would be desirable to ascertain whether denitrification could be promoted in such a system by the alternate use of recharge zones and 'resting' zones.   Indeed, there is scope for adopting more scientific management of recharge sites in the U.K. generally, based possibly on American practice (ref. 12).   In general, recharge sites should preferably have an unsaturated zone of at least 10 m and new sites should have their suitability confirmed by hydrogeological investigation.

23. Outstanding research needs related to effluent recharge include the acquisition of more information on trace organics, as already mentioned, and the optimisation of the actual method of applying the effluent.   Experience in the Southern Water Authority suggests that surface flow across a thin top-soil, as at Ludgershall, is best;   where this is not possible because of ground conditions, the respective merits of trenches, lagoons and French drains need to be investigated, having regard also to the type of effluent applied, the duration of active and resting periods, and other operational factors. Some of these variables are being studied in connection with the changes required at Winchester sewage disposal site.

ACKNOWLEDGEMENTS

This paper is published by permission of Dr. S.C. Warren, Director, WRC Environment, and Mr. P. Lofthouse, Director of Operations, Southern Water Authority.

Any opinions expressed are those of the authors and are not necessarily those of their employers or sponsors.

REFERENCES

1. BAXTER K.M. and CLARK L.   Effluent Recharge.   WRC Technical Report TR199, 1984.

2. MONTGOMERY H.A.C., BEARD M.J. and BAXTER K.M.   Effects of the recharge of sewage effluents upon the quality of Chalk groundwater.   Water Pollution Control, in the press.

3. BEARD M.J. and MONTGOMERY H.A.C.   Survival of bacteria in a sewage effluent discharged to the Chalk.   Water Pollution Control, 1981, 80, 34-41.

4. BAXTER K.M., EDWORTHY K.J., BEARD M.J. and MONTGOMERY H.A.C.   Effects of discharging sewage to the Chalk.   Science of the Total Environment, 1981, 21, 77-83.

5. BAXTER K.M. and EDWORTHY K.J.   The impact of sewage effluent recharge on groundwater in the Chalk for an area in South East England.   Presented at the International Symposium on Artificial Groundwater Recharge, Dortmund, Germany, 1979.

6. TESTER D.J. and HARKER R.J.   Groundwater pollution investigations in the Great Ouse Basin.   Water Pollution Control, 1981, 80, 614-628.

7. BAXTER K.M.   The effects of discharging a primary sewage effluent on the Triassic sandstone aquifer at a site in the English Midlands.   Presented at the first International

Conference on Groundwater Quality Research, Houston, Texas, 1981.

8.   LUCAS J.L. and REEVES G.M.   An investigation into high nitrate in groundwater and land irrigation of sewage.   Progress in Water Technology, 1980, 13, 81-88.

9.   JOSEPH J.B.   The effects of artificial recharge with a primary sewage effluent into alluvial deposits at Cilfynydd, South Wales.   WRC Report ILR689, 1977.

10.   BOUWER H., RICE R.C., LANCE J.C. and GILBERT R.G.   Rapid infiltration research at Flushing Meadows Project, Arizona. Journal of the Water Pollution Control Federation, 1980, 52, 2457-2470.

11.   HARTMAN R.B., LINSTEDT K.D., BENNETT F.R. and CARLSON R.R. Treatment of primary effluent by rapid infiltration.   U.S. EPA Report EPA-600/2 - 80 - 207, 1980.

12.   METCALF AND EDDY INC.   Process Design Manual for Land Treatment of Municipal Wastewater.   U.S. Department of Commerce, National Technical Information Service Report PB-299-655, 1977.

Discussion on Papers 12 and 14

DR L. CLARK, Water Research Centre, Medmenham, UK
Paper 12 from the US and Paper 14 from the UK present the high
and low technology ends of the effluent recharge field. The
US method is to treat the effluent to potable standards at the
surface and then to recharge it through boreholes. This is
very largely a resource conservation measure with only minor
treatment connotations – the recharge is only polishing the
effluent. The paper clearly shows the high level of
pretreatment needed for recharge through boreholes. This also
applies to normal artificial recharge – work done at WRC has
shown that potable water being discharged into supply mains
from a treatment works is capable of clogging a medium
sandstone aquifer. This degree of treatment must make the
water very expensive – one suspects too expensive to be
considered in the UK. In addition to the high cost I would
guess that in the UK the SPDES limits listed for Cedar Creek
would be detrimental to the project. Which water authority
would be prepared for the analytical load needed to prepare
daily maxima for 98 determinands – some of the organics down
to 0.1 microgram per litre. Is this really done at Cedar
Creek?
   The UK method of effluent recharge is to effect minor
treatment before recharge by surface ditches or tile drains.
This is primarily an effluent disposal system which uses the
soil and underlying aquifer as the effluent treatment system.
There is no facility in the UK that I know of where effluent
is recharged through boreholes, although the degree of
treatment in UK facilities prior to recharge is increasing.
The work described in Paper 14 does show that raw effluent
recharge increases the loading of nitrogen to aquifers;
therefore, there is an argument for treatment to remove
nitrogen before surface recharge. This is particularly
important for sandstone aquifers like the Bunter sandstone in
England. Resource conservation so far has had almost no
consideration in the UK but Paper 14 shows that this attitude
cannot continue. In certain catchments the effluent from
recharge sites can be an important factor in maintaining river
baseflows, particularly in drought years.
   It is necessary when evaluating the benefits of effluent

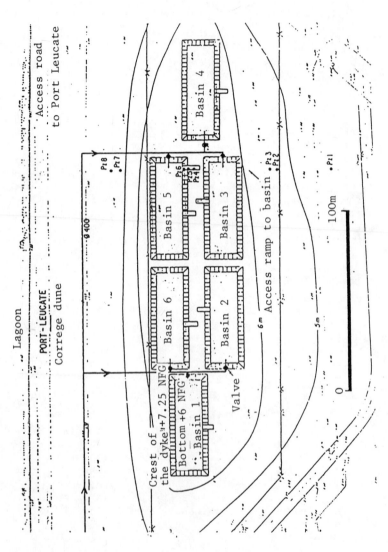

Port Leucate – Layout of the infiltration basins and piezometers

recharge to be able to apply a cost-benefit analysis on the
same bases as for surface sewage treatment works and water
supply systems. Could the authors of Paper 14 give some
indication of the costs per unit volume for effluent recharge
to boreholes in the US and to land in the UK?

Does Dr Avendt consider the high degree of treatment used at
Cedar Creek to be necessary if there were sufficient land
available for surface recharge to be used?

PROFESSOR D. A. OKUN, University of North Carolina, USA
Dr Avendt's experience over a 10-year period provides evidence
that the water quality required for potable reuse of the
recharged effluent is a moving target. How can a massive
capital investment in such potable reuse be justified if the
standards to be met are expected to become more restrictive
over the 20-30 years of the useful life of the facilities?

What percentage of the costs of operation would be
attributable to the monitoring and analyses required for the
potable reuse?

Would a more extensive non-potable reuse programme, similar
to that introduced in Paper 11, possibly eliminate the need
for the very complex, costly and uncertain treatment sequence
illustrated, together with the accompanying monitoring?

MR D. FOUGEIROL, BURGEAP, Paris, France
I was very interested in Paper 14 which reviewed groundwater
recharge of sewage effluent in the UK.

I would like to present results obtained on two pilot wastewater treatment systems developed in France, with the
participation of the financial basis agencies, using the same
process of land infiltration.

Port Leucate (Mediterranean coastal resort)
Six infiltration basins were dug in the sand dunes between the
sea and a lagoon, providing disinfection for pretreated sewage
effluent from a population of 25000. The basins are run
alternately: one is infiltrating while another is drying. In
1980 infiltration tests were made with 1500 m^3/day of raw
effluent. After drying, a black deposit was found inside the
sand filtering layers, which was indicative of reductive
conditions. After 24 hours of contact with the atmosphere the
black deposit was completely removed and a total regeneration
of the filtering capacity of the sand was obtained. This
stage of reoxygenation of the filtering sand after drying is
fundamental to the running of the whole system. The clogging
deposit is then exposed and removed from the basin. The main
part of the extraction takes place in the 6 m of non-saturated
zone between the bottom of the basins and the water table.
Additional extraction takes place in the saturated zone. The
main results with raw effluent are:
- faecal germ concentration is decreased by 4 log units under
  the basin and by 6-7 log units at 50 m distance
- viruses are fully eliminated

229

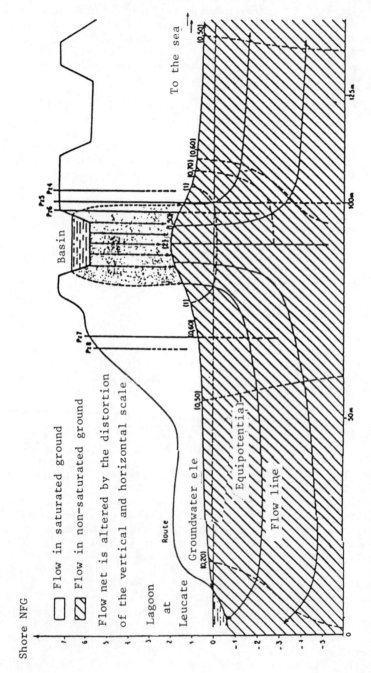

Port Leucate – Flow net under the basin during seepage

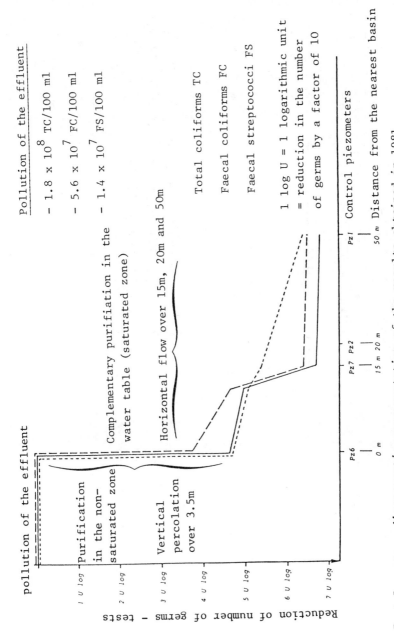

Pollution of the effluent

- 1.8 x 10^8 TC/100 ml
- 5.6 x 10^7 FC/100 ml
- 1.4 x 10^7 FS/100 ml

Complementary purifiation in the water table (saturated zone)

Horizontal flow over 15m, 20m and 50m

Total coliforms TC

Faecal coliforms FC

Faecal streptococci FS

1 log U = 1 logarithmic unit
= reduction in the number
of germs by a factor of 10

Pz1 Control piezometers

50 m Distance from the nearest basin

pollution of the effluent

Purification in the non-saturated zone

Vertical percolation over 3.5m

Reduction of number of germs — tests

Port Leucate — diagrammatic representation of the results obtained in 1981 (gross effluent)

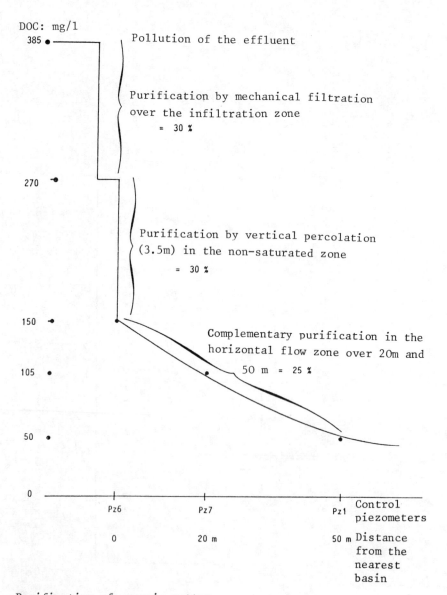

Purification of organic matter

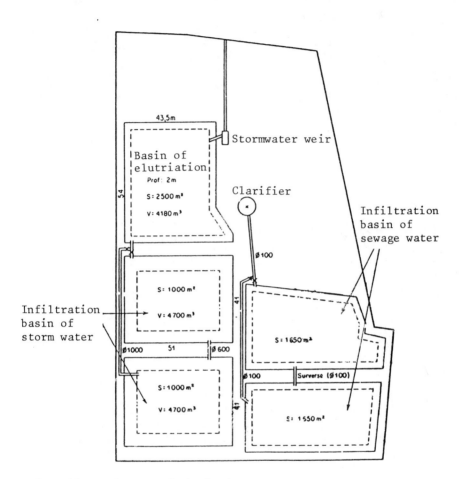

Flesselles - Layout of the basins

- BOD was decreased by 90%
- total nitrogen was decreased by 30-40%
- phosphate was fully eliminated
- the major part of heavy metals remains in the sludge.

In 1983 a physicochemical sewage plant was put into operation. With this system of secondary treated effluent a high infiltration velocity was obtained in the basins, up to 50 m/day, and the bacteriological extraction rate deteriorated.

After testing the extraction rates were found to remain good with infiltration rates less than 10-12 m/day for this particular site.

Flesselles (a village of 2000 people in northern France between Paris and Lille, in the chalk area)

The treatment system is composed of:
- a biological plant which existed prior to the basins
- two infiltration basins for disinfection of pretreated sewage effluent
- a decantation and two infiltration basins for rain water.

The biological station was not operational for two years and on average 300 m^3/day of raw sewage and rain effluent were admitted first into the decantation basins, and then into the infiltration basins.

We obtained the same filtration capacity recovery after 24 hours of exposure to the atmosphere.

With an infiltration rate of 0.2 m/day the results after 0.5 m of percolation in the top sand layer are:
- BOD is lowered by 50%
- reduction of faecal germ concentration is about 1 log unit per 25 cm of sand
- phosphates are reduced by 30%.

These results showed that a layer of sand of 0.5 m is not sufficient for optimal extraction. Tests are also being carried out to estimate the residue to be extracted in the chalk itself as there is 40 m of non-saturated medium on this site.

Results and conclusions
- The infiltrating velocity is the main parameter of the disinfection process in land infiltration
- the alternate running of the basins is fundamental for controlling infiltration velocity, clogging of the sand layer, maintaining the oxygen content of the natural filter and regenerating the filtering capacity
- the non-saturated zone has the same effect as a biological plant; by using fixed bacteria existing in the soil, and natural reoxygenation from the air, additional extraction is obtained for BOD, COD, total nitrogen and phosphate
- disinfection of effluent is very effective for infiltration rates not exceeding 3 m/day, and proves to be a reliable alternative to the classical techniques of sterilization.

The actual research is aimed at the optimization of the

pretreatment unit to meet low infiltration standards and the
optimization of the basin concept, in particular the nature
and thickness of artificial sand or soil filtering layers
where necessary.
  This method is of great value in:
- bacteriological protection of swimming areas or shellfish
  growing along the seashore
- wastewater disinfection before reuse for irrigation in arid
  countries
- protection of limestone aquifers against bacteriological
  contamination.

DR R. J. AVENDT, Paper 12
In response to the comments by Dr Clark, the high degree of
treatment used at Cedar Creek was required by the
demonstration nature of the investigation. Previous studies
performed by the county of Nassau also indicated that a high
degree of tertiary treatment was required to maintain the
recharge well operation. Recharge periods between well
redevelopment were significantly increased during periods of
low concentrations of suspended solids, turbidity and
nutrients. When the recharge operation was through spreading
basins, higher concentrations of solids, turbidity and
nutrients could be tolerated. The comment regarding the high
analytical load reflects a concern of ours. During the
demonstration project, the laboratory operated 24 hours per
day in collecting and analysing the vast number of water
quality standards. Some analyses were automated, but the
majority were conducted by four chemists and six technicians.
This cost of personnel and equipment significantly increased
the overall project budget and must be considered in any
future reuse project of this nature. It must be noted that
the recharge wells and basins are discharged into a sole
aquifer system which is currently the only potable water
supply source for the county residents. Both the high degree
of treatment and extensive monitoring programme were
considered prudent to protect public health.
  The question that Professor Okun raises addresses our
concerns as engineers and scientists working on water reuse
projects. As analytical techniques and public health
investigations advance, the water quality requirements for
reuse projects will become more stringent. The prudent
approach at this time should be to concentrate our efforts on
reuse alternatives that do not involve potable reuse.
Although there may be unique instances where potable reuse is
warranted, the questions concerning fail-safe operation and
real time monitoring have not been resolved and, therefore,
should be considered unknown risks. The costs attributable to
laboratory analysis and monitoring during the demonstration
project were approximately 25 cents per thousand gallons water
recharged or 10% of the operational cost including debt
amortization. The investigation of more cost-effective reuse
options is being continued by the county of Nassau. The

projected shortfall is estimated to be 10% of the permissive
sustained yield by the year 2020. The current water use is
120 gallons per capita per day. A water conservation
programme addressing lawn watering and turf irrigation is
being considered. A programme of industrial reuse/recycle is
also being evaluated. Both of these options should meet the
projected water supply deficit.

DR H. A. C. MONTGOMERY, MR M. J. BEARD and MR K. M. BAXTER,
Paper 14
In replying to Dr Clark we agree that the failure to remove
nitrogen at sites in the Triassic sandstone was disturbing.
However, partial removal of nitrogen could have been caused by
the use of suitable operating procedures with alternation of
active and resting zones as described in Mr Fougeirol's
contribution and in reference 12 of Paper 14.

Mr Fougeirol's report provided welcome confirmation of much
of the work described in the paper.

Costs for recharge of sewage received at the Winchester site
amounted to 1.6 pence per m^3. Where biological treatment was
applied prior to recharge, there was little revenue saving
over costs incurred in the more usual practice of discharge to
watercourse. Typical costs for biologically treated effluent
recharged to chalk and to river were 9 pence per m^3 and
11 pence per m^3 respectively. The work at the Winchester site
in particular has shown that biological treatment of sewage
prior to chalk recharge can be omitted without detriment to
water quality.

15   Health aspects of wastewater reuse

R. G. FEACHEM, BSc, PhD, MICE, FIPHE, MIWES and D. BLUM, BA, MD, MSc,
DPH, Department of Tropical Hygiene, London School of Hygiene and Tropical
Medicine, UK

SYNOPSIS. Extensive information is available on the occurrence
and survival in the environment of most excreted pathogens and
on their removal by various sewage treatment technologies.
Many of these pathogens may be present in sewage in high concentrations, may pass through sewage treatment plants at only
moderately reduced concentrations and may survive for several
days or more on irrigated soil or crops. These facts have been
taken to indicate a potential risk from wastewater re-use and
have led some authorities to set stringent quality standards
for effluents to be used for irrigation. An epidemiological
rather than environmental approach to the problem is advocated
here, in which decisions are based on actual risks rather than
potential risks. The data on actual risks are very scarce at
the present time and do not permit the health consequences of
a particular re-use project at a particular site to be predicted. More epidemiological studies are urgently needed. In the
meantime, a wastewater re-use project should be designed on the
basis of a full analysis of the local circumstances. An epidemiologist should be intimately involved with this design from
the onset, and should also be responsible for setting up the
systems for wastewater quality and health monitoring that are
an essential part of any major re-use project.

INTRODUCTION

1. This paper is brief because most of what can usefully be
said on this subject has already been said. In this symposium
the paper by Pescod and Alka (ref. 1) provides useful data on
the communicable disease risks associated with urban effluent
re-use in developing countries, while the paper by Ridgway
(ref. 2) summarizes the chronic disease hazards associated with
the partial recycling of effluents in the UK. Recently, the
World Bank has commissioned two major reviews that provide ex-
haustive documentation on the health aspects of waste re-use.
The first of these (ref. 3) provides comprehensive information
on the environmental and epidemiological characteristics of the
excreted pathogens. The second, which is so far available only
in summary form (ref. 4), comprises a major review of the health
effects of wastewater irrigation in developing countries and a
detailed consideration of the policy options for wastewater re-

Reuse of sewage effluent. Thomas Telford Ltd, London, 1984

237

use that are available to governments and development agencies. This brief contribution is derived substantially from these two World Bank studies and the reader is referred to them for a more thorough treatment of the issues.

THE DISEASES POTENTIALLY ASSOCIATED WITH WASTEWATER RE-USE

2.   The first distinction to make in considering the diseases that may be associated with wastewater re-use is between the communicable and the non-communicable diseases. The former are related to the microbiological quality of the wastewater while the latter are related to its chemical constituents. In this paper only communicable diseases are considered.

3.   Wastewater is likely to contain at varying concentrations, all pathogens that are excreted by the contributing population. The number of types of pathogens is potentially large, as is the number of diseases which they may cause. Some of these pathogens and diseases are of greater public health importance than others, however, and some pathogens provide useful analogues for groups of closely related pathogens with distinctive epidemiological features. Taking both these factors into account, the sixteen pathogens that should be considered in discussions of wastewater re-use are listed in Table 1. Table 1 also indicates which of these pathogens may be expected to be important everywhere, and which have restricted geographical distributions.

4.   The picture can be further simplified by considering only the broad biological divisions listed in Table 1:
     - the viruses,
     - the bacteria,
     - the protozoa,
     - the nematodes,
     - the trematodes,
     - the cestodes.
This crude categorization is useful for some analyses of wastewater re-use because the pathogens within each biological group share certain important epidemiological and immunological features, as listed in Table 2.

OCCURRENCE AND SURVIVAL OF EXCRETED PATHOGENS IN THE ENVIRONMENT

5.   Information on the occurrence and survival of excreted pathogens in the environment is extensive and has been comprehensively reviewed elsewhere (ref. 3). The only pathogens listed in Table 1 for which our knowledge is seriously deficient are hepatitis A virus, rotavirus and Campylobacter jejuni. In the case of the two viruses, this deficiency results from the extreme difficulty in isolating these organisms from the environment. In the case of Campylobacter jejuni, it is the "newness" of this pathogen, first recognized as a cause of diarrhoea in man in 1973, that accounts for our lack of knowledge.

238

Table 1. Excreted pathogens that should be considered in
          the design of wastewater re-use projects

Biological class	Pathogen	Associated disease	Distribution
Viruses	Poliovirus	poliomyelitis	global
	Hepatitis A virus	infectious hepatitis	global
	Rotavirus	diarrhoea	global
Bacteria	Campylobacter jejuni	diarrhoea	global
	Escherichia coli	diarrhoea	global
	Salmonella spp.	diarrhoea/ typhoid	global
	Shigella spp.	diarrhoea/ dysentery	global
	Vibrio cholerae	diarrhoea	focal
Protozoa	Entamoeba histolytica	diarrhoea/ dysentery	global
	Giardia lamblia	diarrhoea	global
Helminths			
nematodes	Ascaris lumbricoides	roundworm infection	warm climates
	Trichuris trichiura	whipworm infection	warm climates
	Ancylostoma + Necator	hookworm infection	warm climates
trematodes	Schistosoma spp.	schistosom- iasis	focal
	Clonorchis spp.	liver fluke infection	focal
cestodes	Taenia spp.	tapeworm infection	focal

Table 2. Epidemiological and immunological features of the
          biological types of excreted pathogen

Biological class	Transmission	Infectious dose	Does infection confer long-lasting immunity?
Viruses	faecal-oral	Low	Yes
Bacteria	faecal-oral	High	No[a]
Protozoa	faecal-oral	Low	No
Nematodes	development in soil	Low	No
Trematodes	development in aquatic host	Low	No
Cestodes	development in pig or cow	Low	No

Note: the properties listed do not apply to viruses, bacteria
etc. in general but only to those excreted pathogens listed
in Table 1.
a. Except in the case of typhoid fever.

REMOVAL OF EXCRETED PATHOGENS BY SEWAGE TREATMENT PROCESSES
    6.   There is an extensive literature on the fate of excreted
pathogens in conventional sewage treatment works (ref. 3), with
the exception of hepatitis A virus, rotavirus and Campylobacter
jejuni, about which little is known. There is a considerable
but lesser literature on pathogen removal by tertiary treatment
processes and by effluent chlorination. Conventional treatment
plants incorporating primary and secondary treatment stages are
notoriously inefficent systems for the removal of excreted
pathogens. Excreted viruses and bacteria are present in both
the effluent and the sludge in high concentrations and the
heavier protozoal cysts and helminth eggs tend to be concentrated in the sludge, where they may or may not be inactivated
depending on the treatment process.

    7.   A correct selection of high-technology, tertiary
processes, including effluent chlorination, can virtually eliminate pathogens from the effluent and convert it to a microbiological quality approaching that of drinking water. These
technologies are inappropriate in most developing countries,
however, due to high cost and proneness to malfunction.

    8.   The literature on waste stabilization ponds is sufficient
to show that these systems, if well designed and operated, can
eliminate protozoal cysts and helminth eggs and reduce the concentrations of excreted bacteria and viruses to low levels
(ref. 3). These properties, together with their other advan-

tages, of simplicity and relatively low cost, make them a treatment technology of choice in many situations. It must be noted, however, that there is insufficient literature on the removal of pathogens by well designed and operated ponds in tropical and sub-tropical climates and more studies of pond performance in these circumstances are a priority.

EPIDEMIOLOGICAL ASPECTS

9. Much discussion of the health hazards of wastewater re-use has been limited to review of the data on the survival of excreted pathogens in the environment and their removal by sewage treatment processes. These data clearly show that the pathogens listed in Table 1 may be present in sewage in high concentrations, that they may pass through many types of sewage treatment process, and that they may survive in polluted water, in irrigated soil and on irrigated crops for between several days and several months. This ability of excreted pathogens to contaminate fields and crops irrigated with raw or conventionally treated wastewater has led to the assumption that a public health risk exists and to the establishment in some countries of very stringent treatment and quality standards for effluent re-use. While the enforcement of these stringent standards is undoubtedly conservative and prudent, it effectively denies the possibiity of wastewater irrigation to countries that are not economically or technically able to implement the complex tertiary treatment processes required.

10. If, as many speakers at this symposium have argued, there are pressing reasons for wishing to reuse wastewater in many countries, it is necessary to move away from a discussion of the potential health risks posed by the mere presence of pathogens and to examine the actual health risks as reflected by differential disease rates in the population. The actual risks to health associated with particular re-use practices will have numerous determinants and these determinants are likely to be inter-related in a complex and site-specific manner. The determinants of importance may include the type of re-use process, the type of exposure, the properties of the pathogen and disease under consideration and the nature and level of exposure to that pathogen in ways unrelated to wastewater re-use. These potential determinants are briefly discussed below.

Type of re-use

11. The principal types of re-use of sewage effluents are in agriculture and aquaculture. Agricultural and aquacultural re-use may be subdivided according to whether the crop is for human consumption, for animal consumption or is a non-consumable crop. The different implications for health of these types of re-use are discussed in detail elsewhere (ref. 3).

Type of exposure

12. Four groups of persons, potentially exposed in different ways to pathogens transmitted by wastewater re-use, may be

241

identified. Persons consuming a crop or fish fertilized by
wastewater, persons consuming an animal that has consumed a
crop fertilized by wastewater, persons employed at sites where
wastewater re-use is taking place, and persons living near to
such sites.

Properties of the pathogen and disease

13. The risks to health obviously depend upon the.particular
pathogen and disease under consideration. Properties of particular importance are the concentration of the pathogen in the
wastewater, the survival time of the pathogen under various
relevant conditions, the infectious dose of the pathogen, the
life-cycle of the pathogen and the degree to which infection
leads to immunity. The last three of these properties are
summarized in Table 2 and comprehensive information on all of
them is available elsewhere (ref. 3).

14. The relevance of immunity merits further discussion.
If a particular pathogen in a particular community is readily
transmitted to young children, and if infection confers long-
lasting immunity, then a high proportion of older children and
adults may be immune. In such circumstances, wastewater re-use
may cause no extra burden of disease in the community, despite
the fact that the wastewater may contain a high concentration
of the pathogen concerned. Of the excreted pathogens listed in
Tables 1 and 2, this situation theoretically applies mainly to
the excreted viruses.

Other exposures

15. The risk of a disease that is attributable to wastewater
re-use will depend in part on the other exposures to the same
disease that occur in the community. If a particular disease
is transmitted by multiple pathways other than wastewater reuse, wastewater re-use may provide a negligable additional risk.
If, by contrast, the only exposure of a community to a disease
is via vegetables irrigated with wastewater, then the entire
incidence rate of that disease is attributable to the wastewater
re-use practice.

16. If, as will often be the case, the variety and frequency
of other exposures is a function of the level of hygiene in
the community, one might expect that wastewater re-use will
present greater risk in hygienic communities than in less
hygienic communities. Limited epidemiological evidence supports
this view (ref. 4). In other words, as a community becomes more
wealthy and enjoys a more sanitary environment, so the risks of
disease attributable to wastewater re-use may rise. This relationship may be reinforced in the case of certain diseases by
a decreasing level of herd immunity associated with rising
hygiene standards.

17. However, two opposing factors may also operate. First,
as a community becomes more hygienic and infection rates fall,

so the concentration of pathogens in the wastewater will fall
and the level of contamination of the fields and crops may be
reduced. Second, as hygiene improves people are likely to be
more careful in avoiding the risks associated with wastewater
re-use. They may wear protective clothing, wash more frequently,
practice improved kitchen hygiene and in many ways reduce the
risks to their families.

18. These complex and opposing relationships may be illustrated diagramatically as follows:

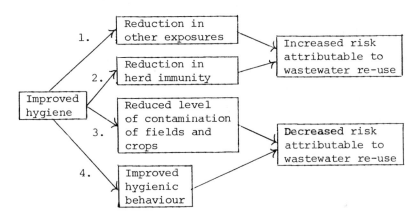

Limited epidemiological evidence suggests (ref. 4) that, in
communities with relatively good hygiene, wastewater re-use
was associated with a risk of nematode infection but not of
viral infection. For the nematode infections, there is negligable immunity (Table 2) and relationship 2 in the diagram
does not apply. The implication is that relationship 1 was
dominant and overshelmed relationships 3 and 4. (It must be
noted that relationship 3 does not apply in communities eating
vegetables irrigated by other people's sewage.) For the
excreted viruses the immunological relationship is potentially
important. The explanation for an absence of risk may be either
that, despite improved hygiene, the level of herd immunity was
sufficient to effectively block any additional risk associated
with wastewater or that relationships 3 or 4 were dominant.
Since relationships 3 or 4 did not appear to prevent a measure-
able risk of nematode infection, the former explanation is more
likely. It may be hypothesized, therefore, that if in a hygienic community wastewater re-use is associated with a risk of
nematode infection but not of enteric virus infection, then
a similar practice in a less hygienic community might be associated with a lesser risk of both types of.infection. In the
absence of more epidemiological evidence, these suggestions
will remain highly speculative.

The epidemiological evidence
19. To illuminate this debate, data on the risks of disease

associated with particular re-use practices in various parts of
the world are required. Such data are very scarce, but have
been recently reviewed in detail (ref. 4). In Table 3 the potential risk of disease caused by excreted viruses (V), bacteria
(B), protozoa (P), nematodes (N), trematodes (T) or cestodes (C)
to various at-risk groups due to different types of agricultural
re-use are indicated. In cases where epidemiological evidence
supports the reality of such a risk, the pathogen is circled,
thus Ⓥ . In cases where epidemiological studies have failed
to find a risk the pathogen is struck through, thus V̸ . Table
4 presents similar information for various types of aquacultural
re-use. The epidemiological evidence on the risks of agricultural re-use is limited and that on aquacultural re-use is
practically non-existent.

Table 3. Theoretical and actual risks of disease caused by
different types of pathogen associated with agricultural re-use of wastewater

Group at risk	Risks from wastewater irrigation of:		
	Crops for humans	Crops for animals	Non-consumable crops
Consumers of crop	Ⓥ, Ⓑ, Ⓟ, Ⓝ		
Consumers of animal		B , Ⓒ	
Workers on re-use site	V̸ , Ⓑ, P , Ⓝ	V̸ , Ⓑ, P , Ⓝ	V̸ , Ⓑ, P , Ⓝ
People living nearby	V̸ , B̸	V̸ , B̸	V̸ , B̸

V = excreted viruses; B = excreted bacteria;
P = excreted protozoa; N = excreted nematodes;
T = excreted trematodes; C = excreted cestodes;
V = potential risk: no epidemiological data;
Ⓥ = risk supported by epidemiological data;
V̸ = risk negated by epidemiological data.

20. More epidemiological studies are perhaps the highest
single research priority in the field of wastewater re-use.
Such studies are methodologically difficult and will require
considerable ingenuity and originality if they are to be conducted at reasonable cost and in a limited time period.

Table 4. Theoretical and actual risks of disease caused by
          different types of pathogen associated with aqua-
          cultural re-use of wastewater

Group at risk	Potential risks from aquacultural re-use for:		
	Crops for humans	Crops for animals	Non-consumable crops
Consumers of crop	V , B , P , (T)		
Consumers of animal		B	
Workers at re-use site	V , B , P , T	V , B , P , T ·	V , B , P , T
People living nearby			

See notes at base of Table 3.

ECONOMIC ASPECTS
21. A formal economic approach to the health risks of wastewater re-use would require that the incremental cost of an improvement in effluent quality or re-use practice be less than
the incremental benefit due to reduced disease. Such an
approach is impossible at the present time because the epidemiological data do not permit a reliable prediction of disease
reduction to be made.

22. Two alternative approaches may, however, be useful.
First, the benefits required to produce a benefit-cost ratio
of 1 can be estimated. For instance, in a particular situation
it might be concluded that the costs of an incremental treatment step would equal the health benefits only if the prevalence
of ascariasis were reduced by 117%. This might cause this
treatment step to be rejected. Second, if epidemiological
judgement can identify a range of treatment and re-use options
with similar health risks, then the costs of these options may
be compared in a cost-effectiveness analysis.

POLICY IMPLICATIONS FOR DEVELOPING COUNTRIES
23. Decision-makers in developing countries cannot wait for
the science of epidemiology to catch up and provide more informed guidance on health risks. Decisions are being taken
now on re-use schemes and these decisions have major implications for public health.

24. The World Bank study on wastewater irrigation (ref. 4)
gives policy guidance based on the limited epidemiological

evidence currently available (Tables 3 and 4). This study concludes from the epidemiological evidence that the risks from
wastewater irrigation are high for excreted nematodes and cestodes, low for excreted viruses and intermediate for excreted
bacteria and protozoa. Six measures to minimize these risks
are identified:

    i.   restrict the type of crops irrigated;
    ii. select irrigation techniques that minimize contamination
        of the crop;
    iii. improve occupational hygiene of the work force;
    iv. disinfection in the home or at central markets of poten-
        tially contaminated crops;
    v.   chemotherapy and/or chemoprophylaxis;
    vi. enhanced wastewater treatment.

Measures i - iv are regarded as potentially effective but difficult to implement in some developing countries. Measure v
is rightly rejected. The study concludes that measure vi is
the most generally applicable and potentially protects all
groups of persons at risk (Table 3).

    25. The World Bank study links the desirability of measure
vi with the epidemiological conclusions previously discussed
and proposes that a treatment technology is required which
achieves:

       - maximum removal of helminth eggs,
       - reasonable reduction in bacterial pathogens,
       - freedom from odour and nuisance.

The study concludes that waste stabilization pond systems, with
anaerobic pre-treatment, 4 main cells and a minimum overall
retention time of 20 days, fulfill these criteria. Waste-
stabilization ponds have a variety of well-documented, additional advantages over conventional treatment plants.

    26. Pescod and Alka, in their paper to this symposium (ref. 1),
emphasize the removal of pathogens, heavy metals and salinity
as the most important criteria for treatment prior to agricultural re-use. They draw attention to the advantages of lime
treatment, over chlorination or reverse osmosis, in fulfilling
these criteria.

CONCLUSIONS
    27. Both the sources cited in the previous section (refs. 1
and 4) draw attention to the need for flexibility and for decision-making based on a full analysis of the local circumstances. Epidemiologists are not yet ready to propose widely
applicable guidelines for health risk estimation in wastewater
re-use schemes. In ten years time this may be possible. In
the meantime it is important that an epidemiologist is involved
in every new wastewater re-use scheme from its very inception.
The role of the epidemiologist is to analyse the health risks
to the community and to work with the civil engineers, agrono-
mists and economists to design a project that will minimize
risks to public health at reasonable cost. The epidemiologist

should also be responsible for designing the wastewater quality and health monitoring systems that must be a part of any major re-use scheme.

REFERENCES

1.   PESCOD M.B. and ALKA U. Urban effluent re-use for agriculture in arid and semi-arid zones. Proceedings of the International Sumposium on Re-use of Sewage Effluent. London, October 30-31, 1984. Institution of Civil Engineers.

2.   RIDGWAY J.W. Re-use of sewage effluent in the UK - an appraisal of health related matters. Proceedings of the International Symposium on Re-use of Sewage Effluent. London, October 30-31, 1984. Institution of Civil Engineers.

3.   FEACHEM R.G., BRADLEY D.J., GARELICK H. and MARA D.D. Sanitation and Disease. Health Aspects of Excreta and Wastewater Management. John Wiley and Sons, Chichester, 1983.

4.   GUNNERSON C.G., SHUVAL H.I. and ARLOSOROFF S. Health effects of wastewater irrigation and their control in developing countries. Proceedings of the Water Re-use Symposium. San Diego, California, August 21-31, 1984. American Water Works Association.

16

Reuse of sewage effluent in the UK–an appraisal of health related matters

J. W. RIDGWAY, BSc, PhD, Water Research Centre, Marlow, Buckinghamshire, UK

SYNOPSIS. Approximately one third of water supplies in the UK are derived from lowland rivers receiving industrial and domestic wastes. Thus some re-use is taking place, with a downstream community drinking water that is, in part, the waste product of an upstream community. In some cases the level of this indirect re-use is high and during recent years concern has developed over the presence in drinking water, as a result of such discharges, of potentially hazardous organic compounds at low concentration. The nature of this concern is discussed and the work being undertaken in the UK to investigate the problem reviewed.

INTRODUCTION

1. Approximately one third of the water supplies in the United Kingdom are dervied from lowland rivers. These rivers, however, are also used for conveying both treated domestic and industrial waste to the sea. As these rivers are mostly short and often receive waste from large populations, the proportion of waste water in the total river flow can be large, especially during dry periods, with at least 7 million consumers receiving water containing on average > 10% sewage effluent. In the Thames, the Lee and the Great Ouse at points of major abstraction the proportion of sewage effluent can at times be as high as 50 to 60% (Table 1)(ref. 1). The use of such rivers for potable supply has been increasing as the limits of availability of water from ground and upland sources were reached and as a result of plans to rationalise numbers of small sources replacing them with single, large, more effectively monitored sources. Thus, in the UK, some re-use is taking place with a downstream community drinking water that is, in part, the waste product of the upstream community, but after man-made and natural processes of purification in the sewage and water treatment works and in the river itself. Thus, re-use, as practised in the UK is described as indirect.

Reuse of sewage effluent. Thomas Telford Ltd, London, 1984

249

Table 1. Proportion of sewage effluent to total river flow at selected water supply abstraction points.

River	Abstraction point	Proportion of effluent (%)	
		Average	Maximum*
Great Ouse	Foxcote	1.9	17.7
Great Ouse	Clapham	6.8	52.0
Great Ouse	Offord	12.2	58.3
Lee	New Gauge	16.5	81.4
Lee	Chingford	20.3	N/A
Thames	Buscot	5.6	51.1
Thames	Swinford	4.6	33.8
Thames	Sunnymeads	11.9	99.8
Thames	Staines	12.1	101.4
Thames	Surbiton	14.4	141.4

*Using river flow exceeded 95% of the time

2. These discharges affect the quality of the receiving water chemically and microbiologically but no precise information is available about the effect on drinking water quality. It is likely, however, that there would be an increase of both non-biodegradable compounds and the metabolic products derived from those substances which are degraded by bacteria.

3. Although such waters are considered wholesome, concern about the presence of organic compounds has developed for a series of reasons. Medical research has shown that environmental factors can have an important effect on the incidence of certain diseases thereby stimulating medical interest in the role of water as a risk factor. There has also been an increasing application to water of advanced analytical techniques which has shown that such waters contain, at very low levels, potentially hazardous organic substances. Concern has been heightened by international interest. A meeting, set up by WHO to discuss possible health effects of waste water re-use, concluded that although data on chemical composition of drinking water derived from polluted sources was essential, this had to be complemented by epidemiological and toxicological studies (ref. 2).

4. Where there is a considerable commitment to using such sources of drinking water, such concerns justify very careful investigation as any reversal of this situation would involve considerable cost and be difficult to justify using currently available data.

STUDY DESIGN

5. To assess this potential problem, the Water Research Centre (WRC), under contract to the UK Department of the Environment, undertook a programme of work to investigate the possibility that consumption of drinking water over a long

period could result in adverse health effects as a consequence of exposure of individuals to harmful chemicals present in the water as a result of re-use. From the outset, the problem was seen to be complex. The potential range of organic compounds in water was wide; their levels were considered to be generally low; the exposure was likely to be long term and the health effects of particular concern, cancers of the gastro-intestinal and urinary tracts, were not new diseases so our search would be for effects super-imposed on existing patterns on disease. Consequently no single approach was likely to result in success and so we combined epidemiological, analytical and toxicological studies so that situations with varying degrees of re-use were compared to determine:

  a) whether there are significant differences in the health of populations, particularly in relation to cancer which can be correlated with a degree of re-use;
  b) the concentration and nature of the substances present in drinking water as a result of re-use;
  c) the relative toxicity of the constituents of the respective river supplies.

6. The epidemiological studies could reveal any health effects recognisable now and provide a baseline for future studies. The assessment of the degree of re-use in different towns could also provide a basis for the selection of sites for analytical and toxicological studies. The analytical studies would provide fundamental data on the substances present in water supplies and might indicate a need for toxicological studies on the identified compounds of unknown toxicity. Finally the toxicological studies of unknown substances could provide a basis for analytical investigations to identify those active substances present.

EPIDEMIOLOGICAL STUDIES

7. The investigation of possible causes of environmentally associated diseases can be complex. This is particularly true of retrospective studies since the number of unknowns can be considerable. When one is looking for an increased incidence of disease which is found in the population anyway, then all the other contributory factors such as diet, occupation, smoking and alcohol consumption must be considered. The actual exposure to the environmental agent in question must also be quantified. Despite these problems, epidemiology can be a useful tool when the right questions are asked since it is looking at a human population and can give a broad measure of the scale of a problem. Before our studies, there had been several retrospective epidemiological studies which had shown evidence of a slight excess mortality associated with the consumption of river water. The relevant literature has been reviewed (ref. 3). The epidemiological studies undertaken by us investigated the possibility of an association between the degree of contamination of water sources and cancer in the UK. The work was undertaken at the Royal Free

Hospital, School of Medicine, under contract to WRC. The programme of work investigated by way of three retrospective studies the possibility of an association between the long term contamination of water sources by sewage effluent and industrial effluent and cancers of the digestive and urinary tracts. In the first two of these studies, cancer mortality and morbidity (disease not necessarily resulting in death) in London were investigated and in the third, national mortality statistics were considered.

8. The initial study was confined to the London area because a large population is supplied from a variety of sources, including some which have for many years contained the highest proportion of sewage effluent in the country (Table 1). The mortality associated with different cancers, was examined for 29 boroughs and districts in the London area for the period 1968 - 1974. Information concerning the source of water supply to each borough was obtained for both the current situation and back to 1926. Throughout, percent dry weather effluent flow/mean river flow was used as an index of re-use in the subsequent analysis, since it could best reflect the historical as well as the current pollution of the rivers in the study. Alternative indices of re-use based on chemical measurements gave an almost identical picture and so would not affect the conclusions of the study.

9. One of the problems encountered when comparing populations (e.g. the London boroughs of this study) is that the groups differ in a number of ways and not just in terms of the variable under study (in this case the source of water). They may differ in terms of socio-economic status, geography and climate. To compensate for such effects (confounding factors), 11 variables, extracted from the 1971 Census data, were used to describe the influence on mortality of social factors. The principal components were:
    a) population density, % overcrowding;
    b) % manual, % semi- or unskilled, mean socio-economic rank;
    c) % lack of amenities, % unemployed, % new Commonwealth, % migration;
    d) % manufacturing industry, % new residents.

10. Throughout mortality for each sex was expressed using the standardised mortality ratio (SMR)* for ages 25 - 74. Mortality data were obtained from the Office of Population Censuses and Surveys for deaths from the following cancers for the seven year period 1968- 1974 in men and women:
    gastro-intestinal, stomach, intestinal, all urinary, bladder, oesophageal, all cancers.
Lung cancer, bronchitis and emphysema were included as a check on the methodology, since these are diseases one would not expect to be related to water quality.

*SMR is the ratio of the actual number of deaths to the number that would be expected from national death rates given the age distribution of the borough's population.

11. The main statistical analysis was based on multiple regression, a technique which enables the effects of several variables to be estimated simultaneously. It provides a means of assessing whether or not any trend in death rates associated with, say, different levels of water re-use is "significant". This significance must remain after allowance has been made for socio-economic variables and for differences in the size of the populations of each of the boroughs being compared.

12. The data was analysed for men and women separately and for each of the cancers. The simple association between SMR and re-use was significant for gastro-intestinal cancers and more strongly so for stomach cancer (Table 2) but in all cases, except male stomach cancer, the socio-economic characteristics accounted for these statistical associations between re-use and mortality. However, in the case of stomach cancer, a weak residual association remained but once the variation in size of boroughs had been taken into account, even this association disappeared.

13. Once this original study (ref.4) was completed, the investigations were extended to include cancer incidence in the London area (covering 14 south London boroughs)(ref.5). Studies of cancer incidence sometimes allow effects to be discerned more easily than in mortality studies since numbers tend to be larger (not all illness results in death) and the latency period, between exposure and onset, is slightly shorter. Again a number of simple associations with re-use were demonstrated but all disappeared with two exceptions, stomach cancer and urinary cancer both in women, when socio-economic factors and variations in borough size were taken into account (Table 2).

Table 2. Summary of significant epidemiological results after adjustment for social factors.

Study	Re-use measure	Cancer/ Group	Significance of association with re-use	
			unweighted	weighted
London mortality	% effluent	stomach/ men	$p < 0.1$	N.S * at 1 in 10
London morbidity	% effluent	stomach/ women	$p < 0.05$	$p < 0.1$
		urinary/ women	N.S. at 1 in 10	$p \simeq 0.08$
National mortality	contributing population	stomach/ men	$p < 0.1$	$p < 0.05$

*N.S. = not significant

14. The national phase of this epidemiological study took advantage of the wider range of water types, especially river derived sources where, unlike London, there is no long-term storage of raw water before treatment. In this study, to minimise the confounding effects of upland water, only towns receiving less than 10% of their supplies from upland catchments were included. In this study, re-use was measured in terms of the population contributing to the river flow upstream of each abstraction point. There were again simple associations with re-use for a number of disease outcomes but with the exception of stomach cancer in men, all disappeared once adjustments were made for social factors and size (Table 2). A full assessment of this study has been recently published (ref.6).

ANALYTICAL STUDIES

15. The second component of the overall programme of work concerned the identification of organic micropollutants in raw and treated water. Before discussing this type of work, it is important to note some of the technical constraints which affect the interpretation of data. Both the general nature of organic matter in water, and the limitations of currently available analytical techniques have an important bearing on the data produced. In general survey analysis, where the determinands are initially unknown, the results are usually at best only semi-quantitative. Also, as a combination of gas chromotography (GC) and mass spectrometry (MS) is usually the only technique which can be applied, the inherent volatility limitation of GC means that at most 20% of the organic carbon present can be analysed. At present, such work can only be considered as a first step towards a comprehensive understanding of the nature of organic compounds in drinking water. Despite these limitations, such analytical data is a vital element in our programme of work but the analytical data must be augmented with epidemiological and particularly toxicological data before comment can be made on the possible health significance of exposure to the detected compounds.

16. As the organic compounds in raw and treated drinking water are usually present at concentrations of < 1 $\mu g/l$, direct analysis of samples is rarely practical. Some form of isolation/concentration process is therefore necessary, e.g. solvent extraction, XAD resin absorption or freeze drying, and although it is possible, using a range of techniques, to extract a very wide range of substances from water, it is usually only practical to apply one or two of these methods at any one time. Thus an element of selectivity is introduced.

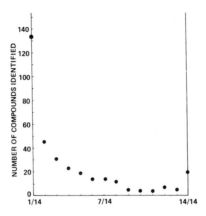

Fig. 1. Frequency of occurrence of compounds identified
in 14 samples of treated drinking water.

17. The detailed results have been reported elsewhere
(ref.7) but in summary the findings were as follows. During
the survey work, samples from a variety of sources were
examined including groundwaters, lowland river waters
including those receiving both sewage and industrial waste,
and upland sources. Some 400 compounds were identified using
GC-MS techniques. The frequencies of occurence are summarised
in Fig. 1. The compound identified most regularly were
hydrocarbons (44%) and halogenated compounds (22%). Many
compounds (41%) were identified only once, implying either
uniqueness to a particular location or a level which only
exceeded the detection limit of the method in one sample.
18. It was difficult to find any relationship between the
type of organic compound identified and the nature of the
source of treated water and no specific association with re-use
was discerned. Although treated river water contained
considerably larger amounts of haloforms (chloroform,
bromodichloromethane, chlorodibromomethane, bromoform) and the
general level of contamination was higher, groundwater and
upland water also contained a complex mixutres of organic
compounds. Although the analytical approach used in this work
was essentially qualitative, levels have been estimated, the
majority at levels below 1 μg/1 (Table 3). Those occurring
at levels of > 1μg/1 were usually haloforms (produced during
treatment) and some fatty acids. Chloroform occasionally
exceeded 100 μg/1 in some drinking waters (ref.8) and
trichlorethylene and tetrachloroethylene could be present in
some groundwaters at higher concentration (> 1μg/1) than in
treated surface water (ref. 9). The semi-quantitative nature

of the analysis made estimates of effectiveness of treatment
and occurrence after treatment tentative. Of the compounds
identified in this work, some 40 halogenated compounds
appeared to have been produced during water treatment
including halomethanes, halogenated aromatics, halogenated
aliphatics, halogenated acetonitriles and halogenated ethers.

Table 3. Number of compounds identified in treated
water versus concentration.

Maximum concentration of compounds (µg/1)	0.01	0.1	1.0	10	100
No. of compounds identified	300+	20	9	4	1

TOXICOLOGICAL STUDIES

19. Although we can identify large numbers of organic
compounds in drinking water, some of which are known to have
carcinogenic properties (Table 4), there are considerable
difficulties to be overcome before a risk to man can be
established and then quantified.

Table 4. Suspect carcinogens in a survey of
14 treated water samples.

Compounds	Frequency of occurrence
Benzene	13/14
Carbon tetrachloride	6/14
Chloroform	14/14
1,2 - dichloroethane	1/14
1,4 - Dioxane	1/14
Hexachloroethane	1/14
Tetrachloroethane	1/14
Trichloroethylene	11/14
Tetrachloroethylene	13/14

For example, the organic compounds present in water occur
at very low levels, typically at microgram level or below,
although exceptionally haloforms and chlorinated alkenes may
be higher. Exposure of consumers to such organics will be at
levels considerably below those at which tumours can be
induced experimentally in tests with mammals (Table 5).

Table 5.   Experimental dose levels for animals and
maximum exposure from drinking water for suspect
carcinogens.

Substance	Lowest experimental dose with tumours (mg/kg/day)	Maximum adult intake from water (mg/kg/day)
Chloroform	60	0.01
Trichloroethylene	870	0.004
Tetrachloroethylene	390	0.00007

Further, although information on toxicity is available for
many organic chemicals found in water, often data on chronic
toxicity is lacking and so it is not always possible to
develop relationships of the type given in Table 5. Therefore,
there remains a considerable element of uncertainty about the
significance of many of the compounds identified in our survey
of water using GC-MS techniques. These uncertainties continue
to be the source of considerable debate (ref. 10)

20. These toxicological problems are heightened when one
appreciates that our knowledge of the organic content of
water is limited to no more than 20% of the total organic
content, i.e. the volatile fraction. The nature of the
remaining 80%, the non-volatile proportion, remains largely
unknown but is the subject of research (ref. 11).

21. Since it is impracticable to detect carcinogenic
effects at the concentrations likely to be encountered in raw
and treated water in experiments using animals, short-term
screening tests to detect mutagenicity are being applied
increasingly in investigations of the potential hazards from
carcinogens in water. These short-term tests for carcinogenic
potential depend on a close correlation between carcinogenicity
and mutagenicity. That is to say that chemicals which can
cause cancer are invariably able to cause mutations (heritable
genetic changes) and vice versa (ref. 12). The most
frequently used assays are those using specially developed
strains of the bacterium Salmonella typhimurium (ref. 13).

22. Initially, we planned to test unconcentrated waters,
i.e. the complex mixture of organics to which consumers are
actually exposed, but our studies showed that this procedure
was beset with many problems and all our testing is now
conducted with either freeze dried or XAD resin extracts
(ref. 14). Tests have been applied to extracts of
concentrated treated water collected from different sites
with different degrees of indirect re-use and attempts are
being made to correlate the observed mutagenic activity with
different characteristics of the water samples. Although
different levels of mutagenic activity were seen in these

samples, it is difficult to determine whether this is related
to the quality of the raw water (a possible function of re-use)
or to differences in water treatment practices at the various
sites. However, the results indicated a positive correlation
between activity in one of the bacterial strains used by us
(TA100) with chlorination and there was a suggestion of an
association between activity in a different strain (TA98).
and re-use.

23. However, before decisions can be made about the
significance of the risk posed by the presence of mutagens
in drinking water, it will first be necessary to confirm
the activity observed in tests with bacteria in assays using
mammalian systems. It will also be necessary to identify the
compound(s) responsible for the activity. This is a far from
simple task since we already know that water contains a very
complex mixture of organics, a large proportion of which is
non-volatile. However, it is possible, by combining
mutagenicity assays with organic analysis, to focus on the
compounds of biological significance. To do this complex
mutagenic extracts are chemically fractionated and successive
fractions are then subjected to further testing and analysis
using a variety of MS techniques (GC, field desorption and
fast atom-bombardment)(ref.15).

DISCUSSION

24. The potential hazards to health associated with the
consumption, over prolonged periods, of drinking water derived
from sources with some degree of indirect use, has been
assessed by us using three main techniques - epidemiology,
organic analysis and toxicology. Each of these components of
the study has provided its own important findings.

25. Although our studies of cancer mortality and morbidity
did reveal a number of simple associations with re-use,
invariabily, once appropriate allowances were made for social
factors and the differences in borough sizes, a number ceased
to be significant whilst the remainder were all reduced in
magnitude. The only significant relationships which remained
where stomach cancer and urinary cancer incidence in women .
in London and stomach cancer mortality in men on a national
basis. Therefore although most of the health indices
considered were not significant, there was some evidence of
small, adverse health effects associated with, but not
necessarily, caused by re-use. Although there is therefore
some degree of consistency between these results and the
findings of some of the epidemiological studies in the USA,
our results are neither clear nor conclusive with regard to
the significance of re-use. Because of the difficulties of
resolving all the confounding factors which might affect an
epidemiological study, further elucidation of the significance
of the re-use factor is unlikely to come from aggregate
population studies of this type. Studies based on individuals
may provide this further guidance.

26. The analytical studies have shown that water contains a complex mixture of organic chemicals. A number of these compounds would be harmful in high concentrations but the concentrations found in water were considerably below those at which tumours have been detected in animal tests. However, we do not know if these low levels are without significance especially where exposure is spread over man's lifetime.

27. Consideration of the results of the analytical work did reveal differences between groundwater, upland and river derived sources both in terms of the concentration and the nature of the organic compounds observed. Nonetheless, high levels of organics could be found in supposedly unpolluted sources, particularly groundwaters, and no clear pattern was established which associated the more toxicologically significant compounds with water derived from sources contaminated with sewage effluent. One important conclusion of the work was the need for more information on the large proportion of non-volatile organic matter present in water.

28. The toxicological studies have been limited both by our incomplete knowledge of the organic content of drinking water and, more importantly, the incomplete chronic toxicity data for the known compounds. Therefore the problems of assessing the hazard that the presence of these compounds might pose to man, especially when the significance of the aqueous route must always been seen in the context of man's overall exposure to organic chemicals in his environment, is very complex.

29. Mutagenicity studies were introduced into the programme in an attempt to overcome some of the constraints of organic analysis and our lack of comprehensive toxicological data. These tests, although widely used, are only just emerging from a developmental stage and it is likely to be some years before decisions regarding the acceptability of organic chemicals in water are likely to based on the results of such assays. Within the objectives of our study, the findings, have not demonstrated any marked association between mutagenicity and re-use.

30. It is unwise at this stage to draw definite conclusions from our work concerning the risks to health associated with indirect re-use of waste water although the risk associated specifically with re-use is likely to be very small. However, our application of short term screening tests for mutagenicity during our assessment of the problem has demonstrated that biological activity, indicative of a carcinogenic potential (as measured in assays using bacteria) may be associated with some aspect of drinking water quality, but not necessarily with re-use. It is likely that the application of such biological assays, especially when used in close association with analytical techniques, will be in an important feature of water quality studies in coming years.

ACKNOWLEDGEMENT

This work was undertaken for and funded by the Department of the Environment and the permission of both the Department and WRC to publish this paper is acknowledged.

REFERENCES
1. WATER RESOURCES BOARD. Water resources in England and Wales. HMSO, London, 1973.
2. WORLD HEALTH ORGANISATION INTERNATIONAL REFERENCE CENTRE FOR COMMUNITY WATER SUPPLY. Health effects relating to direct and indirect re-use of wastewater for human consumption, Report of an international working meeting. Technical Paper 7, WHO-IRC, The Hague, 1975.
3. NATIONAL ACADEMY OF SCIENCES OF THE UNITED STATES. Epidemiological studies of cancer frequency and certain organic constituents of drinking water. NAS, Washington D.C., 1978.
4. BERESFORD, S.A.A. Water re-use and health in the London area. TR 138, WRC, Medmenham, 1980.
5. BERESFORD, S.A.A. Cancer incidence and re-use of drinking water. American Journal of Epidemiology, 1983, 117, 258 - 268.
6. BERESFORD, S.A.A., CARPENTER, L.M. and POWELL, P. Epidemiological studies of water re-use and type of water supply. T.R. 216, WRC, Medmenham, 1984.
7. FIELDING, M., et al. Organic micropollutants in drinking water. TR 159, WRC, Medmenham, 1981.
8. WATER RESEARCH CENTRE. Trihalomethanes in water. WRC, Medmenham, 1980.
9. FIELDING, M., GIBSON, T.M. and JAMES, H.A. Levels of trichloroethylene, tetrachloroethylene and p-dichlorobenzene in groundwaters. Environmental Technology Letters, 1981, 2, 545-550.
10. FAWELL, J.K. and JAMES, H.A. Problems of assessing the toxicological significance of organic micropollutants in drinking water. Proceedings of a conference on "Organic micropollutants in water". Institute of Biology, London, 1981.
11. WATTS, C.D. et al. Nonvolatile organic compounds in treated waters. Environmental Health Perspectives, 1982, 46, 87-99.
12. FORSTER, R. and WILSON, I. The application of mutagenicity testing to drinking water. Journal of the Institution of Water Engineers and Scientists, 1981, 35, 259-274.
13. LOPER, J.C. Mutagenic effect of organic compounds in drinking water. Mutation Research, 1980, 76, 241-268.
14. FORSTER, R. et al. Use of the fluctuation test to detect mutagenic activity in unconcentrated samples of drinking waters in the United Kingdom. In : Water Chlorination. Environmental Impact and Health Effects. Ann Arbor Science, Michigan, 1981.
15. WILCOX, P. and HORTH, H. Microbial mutagenicity testing of water samples. Water Science and Technology. In press.

Discussion on Papers 15 and 16

PROFESSOR M. B. PESCOD, University of Newcastle upon Tyne, UK
Paper 15 has suggested that an overcautious approach to the
reuse of treated effluent would limit application in
developing countries and I strongly support this. Dr Feachem
and Dr Blum are brave to suggest taking risks until more
authoritative information is available but also speculated why
this might be justifiable. I feel that the.symposium should
endorse the authors' suggestion, to encourage more widespread
effluent reuse in developing countries.

Although the recent World Bank report by Gunnerson et al.
(ref. 4 of the paper) has favoured controlling agricultural
reuse by effluent treatment, I feel that crop restriction and
selection of irrigation system to minimize crop contamination
are equally important in arriving at workable reuse projects
with minimum treatment costs. A lot depends on the system
planning, particularly if the downstream reuse is controlled
by the Ministry of Agriculture or if effluent is to be made
freely available to farmers. In the latter case there is
little control of the reuse and maximum treatment is probably
required. In any system design proper management and
monitoring are essential.

The Gunnerson report also came out strongly in support of
wastewater stabilization ponds as providing greater health
protection than conventional secondary sewage treatment
processes. While generally supporting this conclusion, I feel
that more pond operating and performance data need to be
collected in developing countries. In addition, it should be
recognized that ponds are not always feasible as a result of
high land cost or where sand storms are a problem.

I strongly support the authors in their appeal for more
detailed epidemiological studies and recognize the importance
of epidemiological advice in system planning and design. I
have only one specific question to put to the authors: can
they suggest what the risk might be of a cholera outbreak from
irrigating crops with secondary treated effluent and,
considering the cyclic nature of cholera epidemics around the
world, if it might be possible to provide reasonable control
through effluent monitoring for cholera vibrio?

I find it reassuring to hear how the Department of the

Environment is trying to protect our health by funding studies such as those described by Dr Ridgeway in Paper 16. This is a most interesting review of a well-planned and well-executed research project. Unfortunately, or perhaps fortunately, the results are inconclusive indicating the difficulty of attributing the cause of chronic impacts on health to water quality when the concentrations of substances of concern are so low.

The sophisticated equipment and the expertise needed to acquire such data are prohibitive in a developing country and this will be a problem in monitoring effluent reuse projects. Another difficulty in conducting an equivalent epidemiological study in most developing countries is the lack of basic health statistics - the cause of death is not known for much mortality reporting! Nevertheless, the need for conducting reuse-project-related epidemiological studies is apparent and the necessary efforts should now be made to collect data which will help to make future agricultural reuse schemes safe and yet economic.

MR J. A. CROCKETT, Gutteridge, Haskins & Davey Pty Ltd, Melbourne, Australia
In this symposium we have heard from Professor Okun (opening address) and Dr Avendt (Papers 11 and 12) of the stringent quality standards for reuse being applied in affluent America and from Mr Cowan and Mr Johnson (Paper 7) of the equally stringent standards being adopted in oil-rich Middle East states. In discussion several delegates have stated or implied that potable standards must be met for most forms of reuse.

In contrast we have heard from Mr Young (Paper 1) of the significant indirect but actual potable reuse in the UK and from Mr Strom (Paper 3) of the direct reuse for irrigation and industry of disinfected secondary effluent and of raw sewage in Australia, in all cases without apparent detriment to health. Increased production of presumably edible fish by the discharge of raw sewage to the sea is seen by Dr Huggins (Paper 8) as a benefit.

We have also heard excellent but necessarily inconclusive summaries of available information on health risks associated with reuse and their control from Professor Pescod (Paper 6) and in this session from Dr Feachem and Dr Blum (Paper 15) and Dr Ridgway (Paper 16). All of us who visit or operate sewage treatment plants know that the health risk associated with coming into contact with sewage is not as high as long as personal hygiene is good. We also know that many swimming waters are contaminated by sewage and stormwater. In my view simple well-engineered treatment, including detection, followed by sensible reuse represents a minimal health risk and the benefits of this reuse are great.

I agree that we must continue to study the health aspects of reuse and I support the approach proposed in Paper 15. However, I believe we know enough now to say clearly to

politicians and the public that sensible reuse of
appropriately treated sewage has an acceptably low health risk
(especially in dry climates), is beneficial and that we should
fight against unnecessarily stringent conditions.

In case some of you think that we are too relaxed in
Australia I assure you that we too have stringent standards
and similar public reaction to reuse projects to that
described by Dr Avendt in the USA. Most of the reuse schemes
in Australia I showed were introduced by practical water
engineers where the benefits would be great. They are not all
without problems but have been accepted by the benefiting
public. Would anyone here suggest that the scheme should be
abandoned until we can prove them to be totally without risk?

MR J. HENNESSY, Sir Alexander Gibb & Partners, Reading, UK
I agree that there is a health component in irrigated
agriculture projects - but not just in TWW reuse agricultural
projects.

Engineers and medical specialists are in dialogue - e.g.
joint ICID/Tropical Health meeting at the Institution of Civil
Engineers approximately two years ago, and ICID warmly
supports the dialogue.

I support the epidemiological approach but what steps are
now in hand to give a better data base in future years?

With reference to paragraph 24 of Paper 15 why is action
(vi) stressed but not actions (i), (ii) and (iii)? Can the
authors give site-specific examples of how the general
principles are applied in practice?

With reference to Paper 15, and also to Papers 6 and 7,
irrigated agriculture is a rural activity whereas treated
wastewater is an urban phenomenon in the Third World. What is
the perspective for areas served by treated wastewater? Why
do sewage treatment authorities not develop downstream
irrigated areas, under unified management, to grow commercial
crops? Would this not be a suitable development scenario to
show their longer-term benefits to the local community?

PROFESSOR B. DIAMANT, Ahmadu Bello University, Zaria, Nigeria
I wonder why we have to insist on unrestricted irrigation with
wastewater, when the problem can be easily solved by
restricting the irrigated crops to non-edible types, or at
least to types that require cooking prior to consumption. The
wastewater then requires only primary treatment followed by
detention in shallow storage ponds that serve as partial
oxidation ponds. Irrigation should be performed by the local
authority responsible for the central sewerage system, rather
than by private farmers who would need constant control and
supervision.

MR J. P. COWAN, John Taylor & Sons, London, UK
In view of the clear indication in Paper 15 of the extent of
future research in order to provide confident advice what
would the authors identify as the most important actions to be

263

taken by the employer and his advisors to resolve their concerns now?

MR D. FOUGEIROL, BURGEAP, Paris, France

I agree that the health aspect is something we have to be very careful with in places where promotion of wastewater reuse is planned, but in many places in the Third World raw wastewater has been reused for a long time, for example
- In Damascus, irrigation of the Ghouta area is carried out with black effluent downstream of the city. When cholera occurs the people are informed by television announcements that eating of crude vegetables is forbidden.
- In Nonakchott vegetables are irrigated with raw effluent in a situation where the aquifer is brakish.
- In Marrakech, every year the raw effluent is put up for sale.

From this point of view any improvement in the quality of wastewater — which is in some cases the only resource available for irrigation — must be considered as a useful step, even if the treated water does not meet all suitable occidental standards.

DR R. G. FEACHEM and DR D. BLUM, Paper 15

Professor Pescod raises the question of cholera outbreaks in association with the use of secondary treated effluent to irrigate crops. The documented evidence of cholera outbreaks associated with waste reuse is mainly concerned with the use of untreated or very inadequately treated wastes. It is likely, however, that cholera vibrios would pass through many secondary treatment processes and be present in the secondary effluent. There may therefore be a risk of cholera outbreaks associated with the use of such effluents. I think it most unlikely that effluent monitoring for cholera vibrios would prove to be a useful strategy for reducing the risk of cholera outbreaks of this kind. The monitoring of raw sewage for cholera vibrios at various points in the sewer network may well, however, be a useful surveillance tool for identifying the onset and geographical spread of cholera outbreaks in the community.

Mr Hennessy asks about steps now in hand to improve our epidemiological understanding. Several initiatives are being taken. Among them, the World Health Organization is co-ordinating a network of projects in various parts of the world which will study various aspects of the risk associated with reusing nightsoil and wastewater in agriculture and aquaculture. I fully agree with Mr Hennessy that crop restrictions, irrigation technique selection and improved occupational hygiene may well play an important role in attenuating the risks associated with waste reuse.

Mr Cowan asks about immediate action in the absence of good epidemiological information. For projects that are in the planning and design stage, it is essential that epidemiologists be involved in the decision making from the outset. When the project is functioning, surveillance systems, both for the waste quality and for the health of employees at the site, should be established. In addition, the local university or other appropriate institution should be invited to assist in some applied research or monitoring activities concerning health risks. If a small proportion of the project costs could be spent on monitoring and surveillance, considerable progress might be possible.

17 Sewage effluent reuse—economic aspects in project appraisal

R. B. PORTER, BSc(Econ), MIMC and Mrs F. FISHER, BSc, Environmental Resources Limited, London

SYNOPSIS. The approach adopted in the economic appraisal of sewage re-use projects is essentially the same as for other public investment projects. The objective is to select the course of action which maximises the net gain to society through optimal resource allocation. This is achieved by comparing the opportunity costs and benefits associated with each of the various courses of action available and identifying the alternative with the lowest net cost.

INTRODUCTION

1. For many years, sewage effluent was considered a liability which had to be collected, treated and returned to the environment. Increasingly however it is being recognised as a valuable resource; not only does effluent reuse reduce the volume of wastewater requiring disposal but it augments existing water supplies.

2. There are many situations which present opportunities for the non-potable reuse of sewage effluent. These include:

- where freshwater supplies are limited;

- where new freshwater supplies have to be developed at an increasing distance from potential users;

- where a single large water user, or class of user, can tolerate a lower grade of water;

- where receiving water quality requirements necessitate the construction of advanced treatment facilities.

3. Whether effluent should be recovered for re-use in a particular situation obviously requires careful consideration of many factors: technical, health, economic, environmental, financial and institutional. This paper is concerned with just one phase in the project planning cycle - economic evaluation.

Reuse of sewage effluent. Thomas Telford Ltd, London, 1984

267

GENERAL APPROACH

4.   The approach adopted in the economic appraisal of
sewage re-use projects is essentially the same as for other
public investment projects.   The objective is to select the
course of action which maximises the net gain to society
through optimal resource allocation.   This is achieved by
comparing the costs and benefits associated with each of the
various courses of action available and identifying the
alternative with the lowest net cost.

5.   This is easy to state but not necessarily so easy to
achieve; in particular, the quantification and valuation of
the potential benefits associated with environmental projects
can often be problematic.   In this paper we concentrate on
the approach to be adopted to the identification and
valuation of economic costs and benefits.

ECONOMIC COSTS

6.   In economic evaluation, it is necessary to value both
costs and benefits at their real resource values.   This value
can be quite different to that represented by the market
price for a given item.   For instance, the market price may
incorporate subsidies or taxes; these are purely financial
instruments which do not reflect real resource consumption.

7.   The real resource value of a good or service is its
opportunity cost; this is the maximum value which the
resource could earn in an alternative use.   When free market
conditions prevail, the market prices of goods and services
are likely to provide a reasonable approximation to their
opportunity cost.   As such they can be taken as a measure of
real resource value.   But where free market conditions do not
exist and distortions arise through such factors as direct
controls,   fiscal   measures,   monopoly   conditions   etc.,
adjustment to market prices may be necessary.   In practice it
is only worth making such adjustments to major cost items
where significant distortions occur.   Among the items where
cost adjustments are frequently required are land, unskilled
labour, power and imported goods.

8. Land: When land is already owned by a public authority
or is acquired under a compulsory purchase order the price
paid (if any) will often not reflect its value in its most
likely   alternative   use   (existing   or   potential).   An
adjustment is then necessary to reflect the price of similar
land on the open market.   In the absence of a suitable
guideline the value of the land's productivity in its
alternative use would need to be determined.

9   For instance: The land on which the effluent treatment
facilities would be constructed is zoned for urban development.   As the land is in public ownership it would be
transferred to the project at no cost.   In the project

evaluation however the land should be valued at the prevailing market price for building land of the same type.

10. Unskilled labour: In many countries today, despite the existence of considerable unemployment, the wages paid to those in jobs is in excess of the market clearing rate. This arises because wages are determined by institutional (unions) and statutory (minimum wage legislation) factors. The opportunity cost of employing additional labour in terms of output foregone elsewhere in the economy is often less than the actual wage paid. In such cases suitable adjustments need to be made to the cost of unskilled labour.

11. For instance: Unskilled workers who will be employed on the project will be drawn from the agricultural sector where the daily wage rate is under half (say $5) the rate which will actually be paid to project workers ($12). It would therefore be more appropriate in the project evaluation to use a wage rate closer to $5 to reflect the economic cost of the employment.

12. Power: Publicly produced outputs such as power are often state monopolies where prices are sometimes determined by political considerations. The real resource cost can be approximated in such situations by reference to the marginal cost price, that is the cost of providing the next increment of generating capacity.

13. For instance: The price which would be paid for electricity on the project ($0.25 kWh) contains a 50% subsidy provided by the government. This price should therefore be doubled or, alternatively, substitute the cost of producing electricity from the most recent (or planned) generating plant ($0.75 kWh).

14. Imported goods: It is quite common, particularly among developing countries, for the exchange rate of the national currency to be fixed, rather than letting its price be determined in the international money market. This often results in the currency being over-valued, which means that imports cost fewer units of the national currency than would otherwise be the case. Appropriate adjustment is therefore necessary to the price of imported project inputs.

15. For instance: A country where the project will be constructed has artificially fixed its exchange rate to keep down the costs of imports. One dollar exchanges for 2 units of national currency at the official rate and for 2.50 units on the black market. The cost of imports should therefore be adjusted by the ratio of 2.50 to 2.00, i.e. increased by a factor of 1.25.

16. All economic costs which would result from the project should be included, regardless of who incurs them. Economic costs encompass all costs to society which include, for instance, any costs borne by the user in converting his facilities to use reclaimed water. Likewise induced costs should be taken into account, such as increased detergent use when utilising reclaimed water.

Financial costs excluded

17. As economic analysis is concerned with real resource values, purely financial costs are not included. Thus if the project is to be funded wholly or in part with loan finance, principal and interest payments are ignored; the method of financing is irrelevant in economic analysis.

18. Similarly with inflation; general changes in price levels are pecuniary effects and do not represent real resource allocations. All costs should therefore be expressed in constant prices and inflation only needs to be taken into account to adjust for relative price changes between project components; that is, for those items where there is good reason to anticipate that inflation will be higher or lower than the average rate.

Incremental costs only

19. Only those costs which result directly from the project are taken into account. This includes all capital costs (both initial and replacement) and operating costs over the project's life. Costs associated with existing facilities (sunk costs) are excluded.

ECONOMIC BENEFITS

20. The benefits of a water re-use scheme will depend on the use, or combination of uses, to which the water will be put. The potential markets for the reclaimed water must be identified and an estimate made of anticipated usage by each market sector. In theory, reclaimed water will serve almost any non-potable market currently served by fresh water supplies. This includes:

Groundwater Recharge	Water table management
Agricultural	Crop irrigation
	Aquaculture
Industrial	Cooling water
	Boiler feed
	Process water
	Construction uses
Non-potable Urban	Landscape irrigation
	Fire protection
	Toilet flushing

Recreational   Lakes
                                 Marsh enhancement
                                 Fisheries

21. In many cases, the water reclamation scheme is likely to
be single purpose and the identification of the type of
potential benefits therefore relatively straightforward. The
types of potential benefits associated with the uses outlined
above include the following:

-    augmentation of water supply and postponement of the need
     for new freshwater supply development;

-    reduction in pollution loads;

-    reduction in groundwater overdrawing when used for
     groundwater recharge;

-    availability of nutrients when used for irrigation water;

-    development of wetlands and lakes, restoration and
     enhancement of fish and wildlife habitat;

-    creation of scenic waterbodies for recreational and
     aesthetic enjoyment.

22. The several different kinds of benefits which can result
from water reclamation schemes can be categorised into primary
and secondary benefits.

23. <u>Primary benefits</u>: These are the direct benefits
resulting from the project and to which a monetary value can
often be put, e.g. the value of the water replaced. These
benefits may be direct, such as a farmer using reclaimed water
for irrigation or indirect, such as a reduction in dust storms
due to irrigation.

24. <u>Secondary benefits</u>: These are often referred to as
externalities and embrace the spill-over effects resulting
from, or induced by, the project. These benefits which accrue
to the economy are not taken into account in the quantities or
prices of inputs and outputs of the project itself. These
effects are mainly of two kinds: multiplier effects which
result from the expenditure of the incomes generated by the
project and linkage effects which are the increases in income
generated by the additional activity occasioned by a project in
industries which supply its inputs.

25. These benefits are difficult to establish and effects
such as the 'multiplier' might anyhow occur in the absence of
the project through other acts of public policy. Generally
major emphasis should not be given to possible indirect
benefits of this sort.

271

26. **Public benefits:** There are also the possible public benefits such as enhanced environmental quality and aesthetic improvements. The latter is often one of the main justifications for 'beautification' schemes in arid areas. These types of benefits, which are outside the market and have no accepted monetary value, cause considerable problems of quantification and valuation. Various approaches have been devised to help overcome this problem but none is perfect and in many cases only a qualitative assessment can be made.

27. In summary, while all benefits need to be considered, it may only be possible to value the main direct benefits. Other benefits, particularly those relating to general environmental and aesthetic improvements, will be included and a subjective assessment made of their importance.

COMPARING ECONOMIC COSTS AND BENEFITS

28. The comparison of costs and benefits is made by discounting the quantified annual totals of each over the life of the project. The discounted amounts are then summed to establish whether there is a positive benefit or, when several alternatives are being evaluated, which scheme has the highest net benefit. The discounting process is a well known evaluation procedure and is not considered further here, although care should be taken to include replacement costs and terminal values in the annual listing of costs and benefits.

29. **Replacement costs:** Certain items of capital equipment will almost certainly need replacing over the life of the project. Allowance for this replacement should be made in the appropriate year.

30. **Terminal values:** At the end of the project's useful life certain items may still retain an economic value. The discounted value may not be significant in the case of a long life project but for shorter life projects, and for items such as land, which are likely to retain their value, the discounted terminal value should be deducted from total (present value) costs.

31. A further point which should be noted concerns the choice of discount rate to be used. This again involves the opportunity cost concept, in this case what the capital would earn in its best alternative use. This may or may not in practice be similar to the rate of interest on borrowed funds but conceptually it is quite different. It is very likely that the economic and planning authorities will be able to advise on the appropriate rate and the official rate should always be used when one is available.

32. We have noted that it will not be possible to quantify all benefits. For each alternative it is therefore necessary to include a full discussion of the perceived secondary and

public benefits alongside the quantified benefits. As by their nature these benefits cannot be reduced to numerical values, judgement cannot always be excluded in making the final selection.

APPROACHES TO QUANTIFICATION OF DIRECT BENEFITS

33. The general approach to the consideration of the economic costs and benefits in sewage re-use schemes has been described above. Undoubtedly the most difficult area is that of benefit evaluation and in the remainder of this paper we concentrate on considering the different approaches which can be adopted to placing values on potential benefits.

34. When reclaimed water replaces an existing or planned source of alternative supply the economic value of the replaced source can be taken as the benefit measurement. When this approach is not possible the potential benefits must be valued directly. These different approaches to benefit valuation are considered below.

Abundance of fresh water

35. For obvious reasons, a water reclamation scheme with the primary objective of water supply is not likely to hold strong attractions when freshwater supplies are abundant. Where no new water supply facilities are envisaged within a reasonable planning horizon, the marginal cost of existing water supply should be taken for comparing with the cost of reclaimed water. This marginal cost of supply is represented by the costs of operating and maintaining the existing facilities. The price of current water supplies should not be taken; this may incorporate a subsidy element or partly reflect historic costs (i.e. sunk costs).

Scarcity of fresh water

36. It is in situations of scarcity of freshwater supply that water reclamation schemes are most likely to be considered. The approach to the analysis is similar to that just described when water supplies are abundant but with one important difference: the cost of fresh water supply adopted would be the cost of new planned supplies when, as a result of the reclamation scheme, the latter is delayed indefinitely. The marginal cost of supply in this case will normally be higher than the costs of existing supplies as the closest and cheapest sources of supply can logically be expected to be exploited first.

37. If the re-use scheme does not replace an alternative planned supply source but merely postpones its implementation, then the net benefit of the delay should be calculated rather than using the cost of new supplies. This is calculated by subtracting the present value of the costs of the water supply scheme on the delayed schedule from the present value of the costs on the original schedule. This is illustrated overleaf.

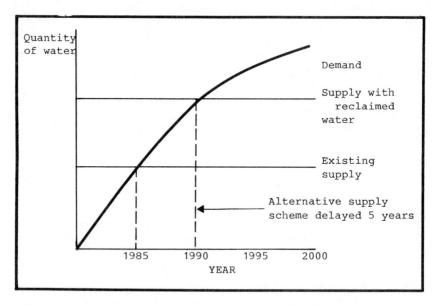

38. The water augmentation scheme originally planned for 1985 would be postponed until 1990 as a result of the water re-use scheme. This delay is a benefit which should be attributed to the re-use scheme.

New uses for reclaimed water
39. It is possible that reclaimed water might be used in an application which does not exist at present and which would not be served in the future in the absence of reclaimed water. This situation could occur when the costs of new freshwater supply exceed the perceived benefits associated with the proposed use. In this situation it would not be justifiable to proceed with a new freshwater supply project and the costs associated with the latter are not therefore relevant in considering the benefits of a water reclamation scheme.

40. The benefits of the reclaimed water have to be determined directly and the approach to this will depend on the particular type of benefit.

41. Crop irrigation: Irrigation is probably the major application of reclaimed sewage effluent. The value of the benefits can be estimated by reference to the sales value of crop production which in turn requires determining the mix of crops which will be grown and anticipated yields.

42. There are further benefits which might also accrue and which would need to be taken into acccount. For instance, if the reclaimed water contains nitrogen and phosphorus, this could result in lower usage of bought-in fertilisers. On the

other hand, the higher levels of solids and salts in reclaimed water could cause premature replacement of irrigation equipment and salt build-up in the soil. An attempt should be made to put a value on these effects and include them in the analysis when they are significant.

43. Recreational use: One of the potential benefits of reclaimed water schemes is recreational use either as part of new or existing projects. In the United States, for example, several projects using reclaimed water have incorporated the creation of reservoirs suitable for fishing and boating.

44. The additional costs of providing facilities for recreational use such as access roads, parking, amenities adjacent to the reservoir etc., should be added to total scheme costs. These are relatively straightforward to identify, quantify and value compared with the potential benefits. Benefit valuation requires an estimate of the value to be placed on the facilities by establishing how many people would use the lake for recreation and what admission price they would be willing to pay if a fee were charged. Different values are likely to be associated with different uses of the lake, e.g. fishing compared with general recreation. The usage of the facilities will need to be estimated over the life of the project taking account of recreational substitutes. The net recreational benefit is determined by multiplying the number of participant days for each use by its value and determining the net present value.

EXAMPLES OF ECONOMIC ANALYSIS OF RE-USE PROJECTS
45. To demonstrate how the principles discussed above would be applied in practice we have selected some examples of typical sewage re-use applications.

Example Case 1
46. In order to meet stricter pollution control regulations and to accommodate anticipated increases in wastewater flows, a community is required to expand its wastewater treatment facility. There are three alternative technical solutions which are feasible; each is based on the same treatment process but involve different methods of effluent disposal.

47. As each alternative involves the same level of treatment, the cost variance between them will be chiefly the difference in transmission costs. This is a function of the distance which is similar in each case. No pumping costs are involved.

48. The selection between the alternatives therefore depends primarily on the different benefits which each scheme provides.

	Effluent Disposal/ Reuse Route	Potential Benefits
Alternative 1	Effluent disposed to local watercourse before flowing to sea	General environmental improvement and improvement in stream quality
Alternative 2	Effluent disposed to surface pond used for irrigation	Reduction in use of groundwater for crop irrigation
Alternative 3	Effluent disposed to a series of ground-water recharge ponds	Reduction in fresh-water currently used for recharge

49. No attempt is made to place a value on Alternative 1. In the case of Alternative 2, as the groundwater displaced is only suitable for crop irrigation (i.e. it is not released for potable use) the benefit value is the marginal cost of groundwater supply (the operation and maintenance costs).

50. In Alternative 3, the saving of potable water valued at its marginal cost represents the opportunity cost of the resources it takes to earn the benefits. As new freshwater supplies are planned it is the cost of the new supplies which should be taken as the benefit measure.

51. A direct quantified comparison will be possible between Alternatives 2 and 3 to determine which provides the highest level of benefits. Alternative 1 can only be considered qualitatively but as the re-use schemes would also provide similar benefits the re-use scheme providing the maximum additional benefits would be selected.

Example Case 2
52. Effluent from a large industrial estate is at present pre-treated (if at all) by individual units before discharging to the sea. It is proposed in future to provide facilities for the collection, central treatment and disposal or re-use of all sanitary and industrial waste waters. The alternative schemes are listed overleaf.

53. The unit economic costs of the re-use schemes are higher than providing for improved pollution control alone: Alternative 1: $0.15/m^3$; Alternative 2: $0.18/m^3$; Alternative 3: $0.35/m^3$. The re-use of effluent for boiler feed imposes considerable additional costs and it is therefore necessary to compare the benefits of each alternative against these costs.

	Effluent Disposal/ Reuse Route	Potential Benefits
Alternative 1	Final effluent disposed to sea	General environ- mental/aesthetic improvement in water quality
Alternative 2	Effluent used for irrigation	Reduction in use of groundwater for crop irrigation
Alternative 3	Effluent used for boiler feed	Replacement of desalinated water for boiler feed

54. The benefits of the reuse alternatives are the savings in the marginal cost of existing water supplies. For groundwater this was $0.20/m^3 and for desalinated water $0.60/m^3; in the case of desalinated water the marginal cost is roughly three times the actual price ($0.22/m^3) paid by users because of government subsidies. It can be seen that although there are high additional treatment costs associated with providing effluent for boiler feed, the savings this provides result in the highest net benefit ($0.60 - $0.35 = $0.25/m^3 compared to $0.20 - $0.18 = $0.02/m^3).

Example Case 3
55. As a result of higher water consumption expected following completion of a major water importation project there is a need for new wastewater treatment facilities. As the city is in a desert area water resources are extremely scarce and it is therefore intended to make use of the treated effluent. Two alternatives are considered:

Alternative 1: Reuse effluent for crop irrigation.

Alternative 2: Reuse effluent for creation of parks and green areas in the city.

56. Th? costs associated with each of the two alternatives should b? relatively straightforward to establish but always remembering to adjust for any distortions in market prices of the sort discussed earlier.

57. The consideration of benefits is less straightforward. In both cases there is no alternative water supply for which the reused water would substitute to provide a measure of benefit evaluation. The potable water supply would not be used for these purposes because of the very high costs of importation.

58. The benefits associated with irrigation use can be determined by estimating the value of crop production. This requires considering the acres of each crop to be planted, crop yields and expected sale value. (It must be remembered that <u>all</u> costs associated with farming and irrigating must be included in the cost schedule, irrespective of who incurs them).

59. The valuation of the benefits resulting from the creation of a green urban environment with the provision of public gardens and parks and other recreational areas is considerably more difficult. One method is to ask people to place a value on the provision of these amenities but it is notoriously difficult to obtain reliable results from such surveys.

60. It is, in effect, not possible to satisfactorily value aesthetic benefits and the use of judgement cannot be avoided in deciding whether to select the provision of cool, green shady city areas in preference to increased agricultural production. This emphasises the point that economic cost benefit analysis can only provide information in a systematic and as far as possible, quantified manner in order to assist decision making. It provides an input into the decision making process; it does not substitute for it.

FINANCIAL EVALUATION

61. In addition to the economic appraisal of re-use schemes it is essential that the selected project be subjected to full financial evaluation. It may be that the economically preferred scheme is not financially viable.

62. Ideally full financial projections should be prepared for the project, including income and expenditure statements and cash flow statements. In preparing these statements there are two areas of critical consideration:

- how, and on what terms, can the investment be funded?

- how can costs be recovered?

63. As this paper is concerned principally with economic aspects of project appraisal it is not appropriate to consider the approach to financial evaluation in detail here. However, it is worth re-emphasising the main points of distinction between the economic and financial approaches.

64. As we have seen, economic evaluation is concerned with the real resource costs to the community of an investment. Financial analysis is much narrower in outlook; its objective is specific, not comparative. It is concerned with the actual market costs of the goods and services to be purchased and with the actual revenues which will be received. For

278

instance, the tax and subsidy elements in costs and revenues which are excluded from economic evaluations are retained in financial calculations. Also, if items such as unskilled labour and foreign exchange have been priced at their opportunity costs, the actual market values must be substituted.

65. Principal and interest payments also receive different treatment. They are excluded entirely from economic evaluation because they do not effect the real resources used; the method of financing is not relevant in economic analysis. In financial analysis, however, methods of funding and cost recovery are vital to the project's financial viability. Similarly depreciation, which is an accounting charge against the use of assets, is important in financial analysis but irrelevant in economic analysis.

66. Finally there is the treatment of inflation. In economic analysis inflation is only taken into account to adjust for relative price changes between project components; in financial analysis allowance for absolute price changes should be made.

CONCLUSION
67. The economic evaluation of re-use projects seeks to determine the net benefit to society of the proposed investment(s). To do this it takes account of all the costs and benefits which will result, not just those represented by the immediate project inputs and outputs. In addition, all costs and benefits are valued in real resource terms (opportunity costs) to reflect the true value to the economy.

68. A financial evaluation using actual market prices, allowing for any subsidies available and taxes payable, will need to be undertaken following the economic assessment. The project which provides the greatest net benefits to society may not be viable in financial terms and it is essential that a financial feasibility analysis is made of the selected scheme.

REFERENCES
1. ERNST and ERNST. Interim Guidelines for Economic and Financial Analyses of Water Reclamation Projects. Prepared for Office of Water Recycling, State (California) Water Resources Control Board, February 1979.
2. CAMP DRESSER and McKEE INC. Guidelines for Water Reuse, March 1980.
3. IRVIN, G. Modern Cost Benefit Methods. Macmillan Press, 1978.

Discussion on Paper 17

MR J. D. PERRET, Datchet, Berkshire, UK
The economic aspects of project appraisal are essentially the
same for all public investment projects. It is a normal
practice for firms to carry out a financial appraisal of
projects that they are contemplating, as they are concerned
with the return they will get on money invested. Such things
as taxes and grants can be very important. It is rare to come
across a full economic appraisal, as given in Paper 17.

The authors point out that apart from economic evaluation it
is necessary to give consideration to many other factors which
cannot be quantified.

I should like to draw attention to two important subjects
that can be very difficult to attribute economic costs and
benefits to, but which are often essential in order to
maximize the net gain to society through optimal resource
allocation.

Firstly there is the subject of human life. Some figures on
the relationship between sanitation and mortality were given
in Paper 2, particularly comparing urban and rural
populations. As a rough guide about half the world's
population, or 2000 million men, women and children, have no
easy access to a safe and adequate water supply. Many more
than 2000 million people lack even the simplest form of
sanitation. 80% of the world's population suffer from
diseases related to poor water supply and sanitation, and many
millions die each year as a consequence.

It follows therefore that in many parts of the world the
treatment of sewage effluent that is already reused, or the
reuse of sewage effluent for industrial and agricultural
purposes, thus releasing clean water for human use, can save
lives - an important net gain to society. What is the
economic value of these lives? Public perception can be
misleading. Single events causing immediate damage are
perceived as worse than a large number of small events causing
less total damage or loss of lives. The current public
response to starvation in Ethiopia illustrates this point.
Far more people die of starvation every year throughout the
world, without mobilizing public sympathy to the same extent.

It has been stated that the consequences of the Ronan Point

Reuse of sewage effluent. Thomas Telford Ltd, London, 1984

281

disaster have cost something like £200 million per life saved, whereas screening the hormone level of all pregnant women in this country, followed by treatment where necessary, would cost less than £50 per still birth prevented. Presumably this puts the value of a human life somewhere between £50 and £200 million which is not much use in an economic appraisal.

The second problem I find in economic appraisal is the question of risk. The risks attached to the reuse of effluent for human consumption are both real and very small. I would like guidance on these two subjects, the value of life, and the incorporation of real, but not easily evaluated, very small risks.

PROFESSOR B. DIAMANT, Ahmadu Bello University, Zaria, Nigeria
The safe disposal of wastewater is the direct responsibility of the local authority concerned as is education and health. Regular economical considerations should therefore not be applied in this case. There is no profit in this disposal, and expenses (borne by the taxpayer) can only be offset by the income of the raised crop. This point should be emphasized when discussing the economics of wastewater irrigation.

MR J. P. COWAN, John Taylor & Sons, London, UK
How would the authors approach the identification of quantifiable benefits arising from the increase/reduction of health risks to the community served by an effluent reuse project?

MR R. B. PORTER and MRS F. FISHER, Paper 17
In reply to Mr Perret, how, and whether, the value of life can be determined adequately is the subject of continuous debate. Because of this it is preferable to concentrate on the quantifiable benefits which might result from a longer and healthier life. Thus the economic worth of an individual's healthier and lengthened life can be calculated by discounting the future stream of changes in the individual's future gross earnings. Alternatively, a person's economic worth can be estimated by calculating the present value of the output he will generate minus the amount that he will consume. But neither of these approaches takes account of the value of his life to him, his family or his friends.

With regard to Professor Diamant's comments, it is always appropriate to consider all the costs and benefits associated with alternative methods of effluent disposal and reuse to determine the option with the lowest net cost to the economy. However, when different bodies are involved in the provision of water supply and effluent disposal, the scheme which maximizes benefits (or provides the lowest net costs) may not always be in the financial interest of all the bodies concerned. The additional costs and benefits may accrue to different authorities or individuals. In such a situation it is to be hoped that a system of charges and/or subsidies can be introduced to provide the financial incentives necessary to implement the economically optional scheme.

In reply to Mr Cowan, to the extent that effluent reuse schemes conform to health guidelines for the particular application being considered, the risk to health should be small. The World Health Organization has suggested suitable treatment processes to meet given health criteria for effluent reuse. The problem arises when the recommended operating procedures and end use applications are not strictly adhered to. This is the real risk to health, and is behavioural in nature rather than technical. Any attempt at health risk assessment is thus in terms of judgements concerning good operating practice. If there are serious reservations concerning the latter it is probable that the effluent reuse scheme should be reconsidered.

18   Advanced treatment processes for the reclamation of water from sewage effluent

C. BOWLER, ARTCS, LRSC, MIWES and S. H. GREENHALGH, BSc, PhD, CEng, MIChemE, Ames Crosta Babcock Ltd, UK

SYNOPSIS.   The different standards to which sewage effluent can be treated are discussed, along with the uses to which the product waters of various standards may be applied. Some of the processes available for the improvement of secondary effluent and the reclamation of water from sewage effluents are described.   In particular, processes and treatment plants designed to produce water of a potable standard are dealt with and a large reclamation plant incorporating reverse osmosis, which is nearing completion in the Middle East, is described.

INTRODUCTION

1. Water is a valuable resource which is essential to life but which is not always available in plentiful supply, even in more temperate climates. It has, therefore, become accepted in many parts of the world that a sewage works discharging biologically treated effluent represents the source of a potential water supply and methods have been developed for the recovery of water from such origins.

2. For the purpose of this paper, processes which further improve the quality of secondary efluents are included in the category of advanced biological, chemical and physical processes.

3. Before discussing some of the methods available for the advanced treatment of sewage effluents, it would seem sensible to briefly review the conventional methods used to treat domestic sewage to a standard generally accepted as being suitable for discharge to an inland watercourse in the United Kingdom i.e. the long standing "Royal Commission Standard" of 20 mg/l Biochemical Oxygen Demand (BOD) and 30 mg/l suspended solids.

4. Domestic sewage is normally passed first through coarse raked screens for the removal of gross solids, rags and general trash followed by passage through one of several available types of grit removal plant. The sewage is then subjected to primary settlement which effects the removal of

approximately 60-65% of the suspended solids present along
with some 30-35% of associated BOD. The effluent resulting
from these stages of treatment is referred to as Primary
Effluent and is often discharged to sea or estural waters
without further treatment.

5. For discharge to inland waterways it is always necessary to remove the BOD and suspended solids to at least the
20:30 values referred to above. This is done by subjecting
the primary effluent to biological oxidation followed by
final settlement. The biological oxidation is achieved
either by passage of the effluent through a biological
filter, i.e. a bed of clinker or plastic media which is
covered by a biological slime, or by aerating the effluent
whilst maintaining in it a dispersion of activated sludge,
i.e. a mass of flocculant micro-organisms. In both these
forms of treatment the biological micro-organisms remove the
soluble organic pollutant matter by utilising it as food for
their metabolism.

6. After separation of the suspended matter in a final
settling tank, the resultant effluent is known as Secondary
Effluent.

7. Should a better quality effluent be required with
respect to suspended solids and BOD, the secondary effluent
can be subjected to tertiary treatment such as granular
media filtration. If complete removal of suspended and
colloidal matter and BOD is required, the secondary effluent
must receive chemical treatment followed by complete physical
separation of solids from liquid. When removal of soluble
components of secondary effluent is necessary, this can
be effected either by advanced biological treatment or
by membrane separation techniques such as Reverse Osmosis,
depending on the substances to be removed.

USES OF RECLAIMED WATER

8. Water recovered from treated sewage effluent may be
put to a variety of uses ranging from irrigation of grassland to replenishment of aquifiers. Sewage effluent may be
treated to a number of different standards; the appropriate
standard being determined by the use for which it is intended. To obtain a water suitable for irrigation of landscapes
it is likely that only tertiary filtration and disinfection
of the secondary sewage effluent will be required. In order
to produce a water suitable for aquifer recharge, chemical
treatment, clarification, filtration, disinfection and
at least partial desalination is likely to be the minimum
treatment.

9. Intermediate water qualities would be required for
other horticultural uses such as plant and crop irrigation,

for secondary domestic use such as toilet flushing and vehicle cleaning and for certain industrial uses such as floor washing and cooling.

10. The main requirement for floor and vehicle washing and toilet flushing is that the water be virtually free from solids which could deposit and "grow" on pipelines or storage tanks and that a sufficient degree of disinfection should have taken place to prevent biological filming.

11. Water for industrial cooling purposes, in addition to meeting the above criteria, is also likely to require either the removal or sequestering of dissolved salts to prevent their build-up and ultimately deposition in the cooling system.

12. Plant and crop irrigation requires water of a more specific composition. Whereas grasses can probably tolerate a total dissolved solids (TDS) content of up to 1500 mg/1, depending on irrigation techniques, plants require water with much lower TDS values. Certain plants cannot tolerate certain specific ions, most of them requiring only low concentrations of sodium and chloride ions. Sodium, in fact, has toxic properties to some plants.

13. Crops also demand low TDS waters, some crops requiring as little as 250 mg/1 TDS and 75 mg/1 chloride ion. Hence desalination is required for such applications.

14. If there is either risk or intention that any of the waters for the uses discussed above will be consumed orally, it is important that the water should be treated to the World Health Organisation Standard for potable supplies, especially with respect to bacteriological purity.

CONVENTIONAL TERTIARY TREATMENT

15. Tertiary treatment normally takes the form of granular media filtration for the purpose of reducing the amount of suspended matter present in secondary sewage effluent. A typical requirement is to reduce the suspended solids from around 30 mg/1 (the value commonly aimed at for secondary effluent) to below 10 mg/1. A degree of BOD removal will also be achieved but this is the BOD associated with the suspended solids rather than soluble BOD which is now removed by filtration. Since each mg/1 of suspended solids in secondary effluent has an associated BOD of approximately 0.5 mg/1, it follows that if 20 mg/1 of suspended solids can be removed by tertiary filtration, then the BOD will simultaneously be reduced by 10 mg/1. Hence a 10:10 (BOD: suspended solids) tertiary efluet is readily achievable by filtration of a 20:30 secondary effluent.

16. When secondary effluent is filtered in a downwards direction, the suspended matter retained on the filter media is removed intermittently by passing a flow of clean water upwards through the media bed, usually after or whilst blowing air through the bed.

17. The first type of filter used for this purpose employed a bed of fairly coarse sand as the filter media.

18. On backwashing sand media, reclassification of the sand occurs so that the coarser, heavier grains fall to the bottom and the smaller, lighter grains occupy the upper layer, with the result that surface filtration predominates and the filter chokes fairly quickly.

19. A subsequent development was the use of filters through which the effluent passes in an upflow direction. This overcomes the problem of classification but the filters sometimes tend to suffer breakthrough of solids. Nevertheless many installations of this type operate successfully.

20. Cross flow or radial flow filters have also been developed and are available. These normally operate on a continuous basis whereby the sand media slowly falls down a column across which the secondary effluent passes. As the suspended solids are removed the sand gets dirtier until it reaches the bottom from where it is recycled by air lift to a cleaning system at the top of the column. The clean sand then falls to the top of the bed to recommence its downward travel.

21. Another development in downflow filtration involved the introduction of a layer of coarser but lighter media, such as anthracite, on top of a bed of sand. The coarse anthracite has a large solids retention capacity, thereby permitting longer filter runs. The lower layer of sand is utilised to polish out the finer solids which are not retained by the anthracite. This type if filter is commonly in use today.

22. A more recent innovation is to revert to the deep bed of coarse sand and to backwash with a combination of air and water applied simultaneously. This creates great turbulance within the sand bed and prevents the reclassification into coarse and fine grades which occurs when water and air are used seperately. This enables filtration to proceed in greater depth.

ADVANCED BIOLOGICAL TREATMENT

23. Conventional treatment of sewage is normally designed to achieve a 30:20 suspended solids: BOD effluent. Such a treatment process is not designed to remove ammoniacal

nitrogen or phosphorus; elements which encourage algal growths in receiving water courses.

24. The use of extended aeration in the biological treatment stage provides more complete oxidation of the organic compounds present and also produces more granular activated sludge with enhanced settling properties by effecting a degree of aerobic stabilisation. Consequently a secondary effluent of considerably better standard than 30:20 can be achieved; often a 15:10 standard or even 10:10 standard is obtained. Simultaneously with the oxidation of organic matter, the ammonia is oxidised to nitrate and, when alternate aerobic and anaerobic (anoxic) stages are provided as exist in the oxidation ditch process or in the "Bardenpho" Process, the nitrate produced in the aerobic stage will be released as nitrogen gas to the atmosphere.

25. It is now possible to design plants which consistently produce secondary effluents of the following typical analysis COD <70 mg/l, BOD <10 mg/l, SS <10 mg/l, $NH_3(N)$ <1 mg/l, TKN <5 mg/l, $PO_4(P)$ <1 mg/l, $NO_3(N)$ <5 mg/l.

26. Effluent qualities better than 20:30 can be achieved by increasing the retention time in the aeration tanks and designing secondary clarifiers with low overflow rates. Nitrification (the oxidation of ammoniacal nitrogen to nitrate) also occurs in some of these plants and dentrification (the reduction of nitrate to nitrogen) is now fairly commonplace.

27. Phosphorus removal by chemical precipitation using iron or aluminium salts has been practised for some time. Biological phosphorus removal is also now being employed.

BIOLOGICAL REMOVAL OF NITROGEN AND PHOSPHORUS

28. It has been shown in several full scale plants, that 90% removal of nitrogen and phosphorus compounds from sewage is achievable by modification of the activated sludge process. It appears that to achieve high enough COD : TKN ratios for the maximum possible nitrogen removal, primary sedimentation must be omitted and only screened and degritted sewage should be fed directly to the secondary treatment stage. If 50 - 70% nitrogen removal is adequate then primary sedimentation can be used if preferred. For successful biological phosphorus removal there must be near zero concentration of nitrate in the return sludge. This means that either the process must not include nitrification or alternatively almost complete dentrification must be achieved.

29. Fig 1 shows the process used for about 50% nitrogen removal and this can be used for either raw or settled sewage. The primary anoxic zone should give 1-2 hours

nominal retention and mixing with either submerged turbines or pumps giving a mixing energy of 7 Watts/m^3. The main aeration compartment must have a sludge age sufficient for nitrification.

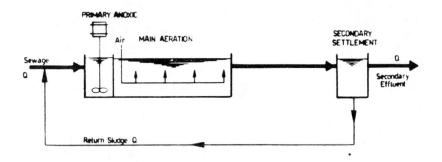

Fig. 1. Typical scheme for 50% N removal

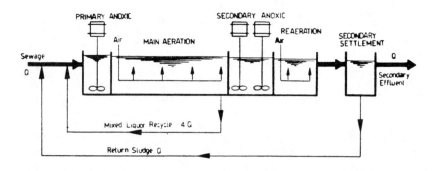

Fig. 2. Typical scheme for 90% N removal

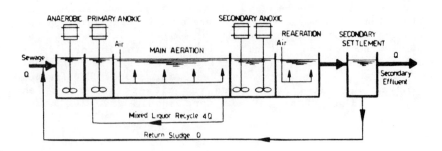

Fig. 3. Typical scheme for 90% N and P removal

30. Fig 2 shows the process used for 90% nitrogen removal which is similar to Fig 1 but with the additions of a secondary anoxic reactor of 2-4 hours nominal retention, a reaeration chamber of 1 hour retention with a residual oxygen concentration of 3-4 mg/l and a mixed liquor recycle from the discharge end of the main aeration compartment to the primary anoxic reactor of typically 4 times the average sewage flow.

31. Fig 3 shows the addition required to Fig 2 to additionally achieve phosphorus removal. This is an anaerobic reactor giving 1 hour nominal retention.

32. It is thought that the primary anoxic reactor removes nitrate from the return sludge and recycled mixed liquor by virtue of facultative bacteria which use simple carbon compounds in the sewage as substrate and nitrate as a source of oxygen. The nitrate is reduced to nitrogen which is stripped from solution in the main aeration compartment.

33. Phosphorus compounds in sewage are converted in the sewer and treatment plant to orthophosphate by natural chemical and biological processes. If activated sludge is held in contact with sewage under anaerobic conditions, orthophosphate is released into solution by specialist micro-organisms such as acinetobacter but all orthophosphate is rapidly taken up into the cells in the subsequent aeration stage. The reaeration chamber is essential as it allows nitrification of the ammonia and reabsorption of phosphate released in the secondary anoxic reactor, strips out nitogen gas and maintains the secondary settlement tank in an aerobic state thereby preventing further phosphate and ammonia release.

ADVANCED CHEMICAL TREATMENT

34. When a clear, low coloured, suspended solids free water is required, chemical treatment is invariably necessary. This will probably be followed by clarification and certainly by filtration. A disinfection stage is usually incorporated.

35. Biologically treated sewage effluent commonly contains some colouring, colloidal and suspended matter, all of which it is desirable to remove to produce a better quality water. This removal can be achieved by flocculating with a primary coagulant such as aluminium sulphate, ferric sulphate or ferric chloride. The optimum dose of one of these chemicals may be such that either an acid or an alkali may be additionally required to adjust the pH to the appropriate value for flocculation, usually 6.0 to 6.5 in the case of aluminium sulphate and 7.0 to 8.5 in the case of ferric salts.

36. The quantity of solids produced from this reaction is likely to be too great to apply directly to sand or dual media filters and, in this event, a clarification stage will be required. This may take the form of dissolved air flotation or sedimentation.

37. Dissolved air flotation operates by saturating a portion of the clarified water with air under pressure and recycling the saturated water to the inlet of a flotation cell. The sudden drop in pressure of this saturated re-cycle on entering the flotation cell causes the air to be released from solution in the form of very small bubbles which rise to the liquid surface. As they do so, they become attached to the flocculant suspended particles thereby lifting them to the surface where they collect as a blanket and are removed by scraping (fig 4).

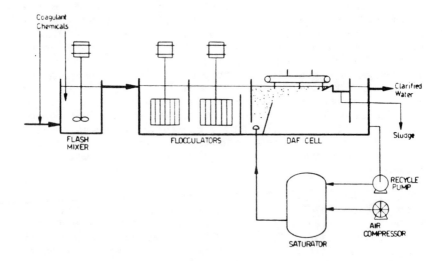

Fig. 4. Dissolved air flotation

38. Sedimentation may be carried out in any of a number of differently designed tanks but which will operate either with a floc blanket or by simple settlement. In the former the chemically dosed water enters the sedimentation tank at the bottom and passes vertically upwards through a suspension of the aluminium or ferric floc. Incoming solids are entrained in the floc blanket and the clarified water is collected in troughs at the surface. In a settlement tank the water will often be flocculated in an inner portion of

the tank prior to passing into the outer zone where the suspended solids (flocs) settle to the tank floor, from where they are removed as sludge, whilst the clarified water rises to surface collection troughs.

39. In some cases it is desirable for the flocculation reaction to take place at a pH value of 10.5 - 11.0. This is achieved by dosing with lime. At this pH a lime softening reaction takes place which results in the precipitation of calcium carbonate and magnesium hydroxide. The latter acts as a primary coagulant, although it will usually be necessary to add additional coagulant in the shape of aluminium or ferric salts or a coagulant aid (eg. a polyelectrolyte).

40. For this reaction it is advisable to use a solids recirculation type of sedimentation tank in which the incoming chemically dosed water is brought into contact with pre-precipitated calcium and magnesium compounds to improve the speed and efficiency of the reaction. Solids/liquid separation is effected in the outer zone of the clarifier as described previously.

41. Advantages of this type of chemical treatment are that it provides a high degree of bacterial and virus kill, produces very good flocculation, heavy flocs with good settling properties and a good quality clarified water. At this elevated pH the water can be applied directly to an ammonia stripping tower, if necessary. This is rarely a requirement however since ammonia can more conveniently and economically be removed in the biological sewage treatment process or by reverse osmosis, if this is included. Alternatively, if ammonia does exist in the effluent, it may be desirable to keep it there, if the treated water is to be used for irrigation, to obtain the nutrient value of the ammonium salt.

42. After settlement of the solids, the clarified water is passed to sand or dual media filters for removal of the fine solids which are too small and light to settle in the clarifiers. It is necessary, however, to depress the pH of the clarified water before applying it to the filters to precipitate the residual coagulant and to prevent deposition of calcium carbonate on the filter media which would cause severe chokage. The calcium carbonate tends to grow round the media grains rather than deposit on the surface, therefore making it extremely difficult to clean by backwashing.

43. Sulphuric acid or alternatively hydrochloric acid is normally added to the clarified water to depress the pH value, however, carbon dioxide has been used for this purpose.

44. The carbon dioxide may be obtained by burning propane

or butane or by recalcining the sludge removed from the clarifiers in which the lime has reacted with the temporary calcium hardness to cause the precipitation of calcium carbonate. This calcium carbonate sludge may be dewatered on a vacuum filter, plate type filter press or equivalent before feeding to a furnace where it is recalcined, i.e. converted into quicklime (CaO) with the evolution of carbon dioxide. The quicklime can be slaked to convert it back into hydrated lime for re-use in the treatment process, whilst the carbon dioxide can be injected into the clarified water to reduce its pH value. Whilst this process may offer economic advantages by effecting a saving in lime and sulphuric acid, it also suffers from some disadvantages. Firstly the sludge from the softening process will contain compounds other than calcium carbonate such as magnesium and iron salts and organic material. The inorganic salts will build up in the recycling process unless controlled and therefore a certain percentage of the sludge must be continuously wasted. Secondly the furnace requires a high amount of maintenance and thus requires a relatively high percentage of down time.

45. The acidification of the clarified water with the carbon dioxide is also prone to some minor complications. The surplus calcium hydroxide solution (lime) in the clarified water will react with the carbon dioxide to precipitate calcium carbonate which provides an additional suspended solids load to the filters and the pH is rarely reduced adequately with carbon dioxide alone and supplementary acid dosing in the form of sulphuric or hydrochloric acid will probably be needed to reduce the pH.

46. Therefore, whilst reagent costs may well be higher when using lime and acid than when regenerating lime and carbon dioxide from the softening sludge, capital and maintenance costs associated with the recalcination and recarbonation plant will be higher than those of the chemical dosing plants.

47. The filters used for 'polishing' the clarified water in this type of process are of the water treatment type utilising finer media than is used in the tertiary type filters described earlier. However, once again the media may be single grade silica sand or sand with a layer of anthracite on top. The big advantage of the single sand media type is that thorough cleaning can be carried out by combined air and water, whereas sand/anthracite filters can not be cleaned by this method as anthracite would be lost over the washout weir and the media would mix.

Disinfection

48. The water will additionally require disinfection and this may be carried out by dosing with chlorine (either from

gas cylinders or from sodium hypochlorite) or with ozone. Other disinfectants such as chlorine dioxide may be used but have been applied only to a limited degree so far. The disinfectant may be added to the filtered water but there is something to be gained from adding it to clarified water feeding the filters in order to prevent algae or other biological growths occuring in the filters.

Additional Treatment

49. The chemically treated, clarified, filtered and disinfected water will then be clear, free from suspended matter and should be low coloured. (In this condition it will be suitable for most secondary domestic uses and, providing the TDS is not greater than 1500 mg/l, for irrigation of most grasses, trees and shrubs used in landscaping).

50. If, however, the water still contains undesirable chemicals in solution such as chlorinated organics or chemical colouring matter, it may be necessary to either pass it through a bed of granular activated carbon or alternatively dose powered activated carbon into the feed to the clarifier to adsorb the offending polluting matter. If, on the other hand, the filtered water contains too great a concentration of dissolved solids for the intended uses, it will require at least partial desalination. The most popular and economical method of achieving this is by reverse osmosis.

DESALINATION BY REVERSE OSMOSIS

51. The process and technology of reverse osmosis has been well documented in very many papers and it is not the intention to repeat that extent of detail here. Suffice to say that in very simple terms reverse osmosis is the application of a water with a high concentration of dissolved solids to a semi-permeable membrane at a pressure in excess of the osmotic pressure of the water, with the resulting passage of desalted water through the membrane whilst the majority of the dissolved salts are retained by the membrane and discharged as a concentration of the feed water. Typical component removal efficiencies are shown in table 1.

52. Using this process, a water containing several thousand milligrams per litre of dissolved solids can be converted into one containing less than 500 mg/l of dissolved solids, the World Health Organisation highest desirable level for potable water.

Table 1. Typical reduction of feed solutes in an R.O. plant.

Constituent	Feed Concentration (mg/l)		Permeate Concentration (mg/l)		
	Average Value	Range	Average Value	Range	Percent Reduction
Calcium Ca	119	92–156	1.56	1–2.4	98.7
Magnesium Mg	49.1	40–67	2.15	1.5–2.8	95.6
Sodium Na	257	195–294	33.7	28–46	86.9
Potassium K	15	13.7–18	2.35	2.05–2.8	84.3
Ammonia NH_3	7.1	4–11	1.25	0.7–1.6	82.4
Total Hardness $CaCO_3$	97.3	470–510	13	12–14	97.3
Bicarbonate HCO_3	63.8	42.9–105	11.3	8.9–12.8	82.3
Sulphate SO_4	423	339–439	1	0.5–5.1	99.8
Chloride Cl	407	355–466	38.5	33.6–49	90.5
Nitrate NO_3	25.9	18.5–32	9.58	6.8–11.7	63.0
Fluoride F	5.08	3.7–6.6	0.36	0.2–0.61	92.9
Silica SiO_2	11.3	4–16	1.5	1–2	86.7
Phosphate PO_4	25	17–32	0.05	0.02–0.08	99.8
BOD	1.38	0–4.3	0		100
COD	31.9	20.8–38	4.07	0–9.4	87.2

53. In order for this process to work efficiently, however, it is necessary to treat the membranes with respect.
This involves ensuring that the water delivered to them is
at a temperature and pH value compatible with their maximum
life and operating efficiency and that it has been pretreated to prevent precipitation of salts on the membrane
when it is concentrated. It should also be free from
colloidal organic matter or residual coagulant from the
pretreatment process which would foul the membrane. Control
of the residual coagulant can be effected by ensuring that
the pH value of the water is adjusted to the optimum one for
complete precipitation of residual iron or aluminium prior
to filtration.

54. The removal of organic colloidal matter is dependent
on selection of the optimum treatment process for a particular secondary sewage effluent and management of that
process to ensure that the treated water is always of a
standard to permit trouble free operation of the reverse
osmosis plant.

55. It is also worth mentioning that not all secondary
sewage effluents are amenable to the same treatment process.
Some are efficiently treated with the liming process whilst
others are better .treated by flocculation with aluminium

sulphate. The alkalinity of the effluent is often an influencing factor on the optimum process. Effluents with high alkalinities are often better treated by flocculation with lime at high pH values whilst low alkalinity effluents are often amenable to coagulation with alum or perhaps ferric salts at fairly neutral pH values. Some effluents require treatment with activated carbon, others have the need for oxidation with ozone for example. Trade effluent discharges into the sewers can be instrumental in creating these requirements. Treatability tests in the laboratory or with a pilot plant at the site of the sewage treatment works should ideally be carried out whenever the opportunity exists.

Jeddah Water Reclamation Plant

56. An example of the advantage to be gained from examination of the effluent can be illustrated by the author's personal experience. Ames Crosta Babcock are currently completing the installation of a water reclamation plant at a sewage treatment works serving the city of Jeddah in Saudi Arabia. At the tender submission stage for this scheme, the original design included lime softening of the secondary sewage effluent with polyelectrolyte coagulant, settlement of the precipitated solids, dual media filtration, chlorination and dechlorination with sulphur dioxide in order to render the water suitable for feed to a reverse osmosis desalination plant.

57. It was also envisaged that the sludge from the softening process would be thickened, dewatered and fed to a lime furnace for recalcination and re-use with the carbon dioxide generated by this process being injected into the clarified water for pH control as described previously.

58. However, advantage was taken of the availability of the secondary sewage effluent and laboratory scale treatability tests were carried out. These quickly established that the effluent contained an inhibitor to the lime softening process with the result that far less calcium carbonate would be precipitated than would be required for economical operation of the recalcination process. This therefore had to be replaced with hydrated lime and acid dosing facilities.

59. The laboratory tests additionally indicated that the polyelectrolyte was far less suitable as a coagulant than ferric chloride or sodium aluminate and that dual media filtration showed no benefit over single media sand filtration. Advantage was taken of this fact to change to sand filters with the benefit of combined air and water washing.

60. The major discovery resulting from the laboratory tests, however, was that the proposed method of treatment

failed to produce a water of the quality demanded by reverse osmosis membranes as measured by the Silt Density Index or Fouling Index (SDI or FI); these being a measurement of the membrane fouling potential of a water. Further tests were carried out using activated carbon and using ozone. The latter was found to give a good improvement when dosed prior to filtration instead of chlorine but an even better result when dosed into the secondary effluent before the lime flocculation stage. This had the beneficial effect of reducing the detergent concentration of the effluent in the ozone contact chamber by removal of the foam produced with the surplus ozonated air and also of producing less odourous solids with better settling properties at the subsequent flocculation and settlement stage.

61. Ozone treatment was consequently incorporated into the scheme and the water reclamation plant consists of aerated equalisation basins to balance the flow and prevent the contents becoming septic, high lift pumping to an ozone tank with associated ozone generation and injection plant, lime and ferric chloride dosing, reactor clarifiers with associated sludge thickening, pH correction by acid dosing, sand filtration, sequesterant dosing (to prevent the deposition of calcium salts in the RO membrane), reverse osmosis, final pH correction with lime and disinfection with sodium hypochlorite. Facilities exist to dose the clarified water with sodium hypochlorite and the reverse osmosis feed water with sodium sulphite for dechlorination if necessary.

62. The plant will produce 30000 m^3/d of water with a TDS of less than 1000 mg/l from a secondary efluent with a TDS up to 5800 mg/l. The quality of the product water will comply with the World Health Organisation maximum permissible limits for drinking water.

19 Process selection and design aspects relating to advanced treatment of sewage effluents

L. R. J. VAN VUUREN, MSc, DSc, and B. M. VAN VLIET, MSc, PhD, National Institute for Water Research, Pretoria, South Africa

SYNOPSIS. Research conducted by the National Institute for Water Research on the reuse of sewage effluent extends over a period of some 20 years. This research was initiated at Windhoek, SWA/Namibia, and culminated in the design and construction of a 4.5 Mℓ/d plant in 1969. This plant plays an important role in meeting the City's expanding water demand. The design of the Windhoek plant was progressively adjusted to incorporate new research findings from laboratory and pilot-plant studies conducted elsewhere in South Africa. Unit processes, which were extensively researched, include high lime treatment, ammonia stripping, quality equalisation, breakpoint chlorination, ozonation, and activated carbon adsorption as well as combined physical-chemical and biological methods. Other associated research projects which include dissolved air flotation, biological nutrient removal, stabilisation, chlorine dioxide, health related aspects and epidemiological surveys, complement the overall research programme. Current research is directed mainly towards activated carbon adsorption and regeneration, ozonation, the use of chlorine dioxide and the control of organohalogen precursors. The success achieved over the years with the development of reuse technology has provoked great interest from water supply authorities and consultants in many parts of the country and several large-scale schemes are already being considered.

INTRODUCTION

1. In southern Africa the reclamation of purified sewage effluents has since the early sixties received a great deal of research effort. Pilot-scale studies at Windhoek, South West Africa, using maturation pond effluent as source resulted in the commissioning of the full-scale Windhoek water reclamation plant (WRP) in 1969. Further pilot-scale investigations at Pretoria using humus tank effluent (HTE) led to the construction of the 4.5 Mℓ/d Stander water reclamation plant (SRP) during 1970. The SRP which served as a research/demonstration facility generated valuable design and process criteria which were progressively implemented at the Windhoek plant.

2. Pilot-scale treatment of raw wastewaters using integrated physical-chemical-biological techniques has been in

Reuse of sewage effluent. Thomas Telford Ltd, London, 1984

299

progress since 1974 and this method has also been tested on larger pilot scale (Athlone plant) for reclaiming effluents which contain a high component of industrial effluent. Although the integrated process approach has not as yet found full-scale application, it is considered to have great potential for the future.

3. In addition to process development work extensive back-up research was conducted which focussed on health related aspects. Development of biomonitoring techniques[1], conductance of epidemiological surveys[2] and monitoring of micro-pollutants[3] formed part of these investigations in which many researchers from the National Institute for Water Research (NIWR), health departments, local authorities and consulting firms participated.

4. Particular attention has been given to biological phosphate and nitrate removal methods and a large number of nutrient removal plants are currently in operation in many parts of the country[4]. When this type of plant was later incorporated at the Windhoek sewage works it had a beneficial effect on the reclamation process in that a much superior quality of effluent became available for further treatment. A spin-off of this research has found practical application for reuse in industry and also for the treatment of eutrophied surface waters.

5. The current research programme is directed mainly at the removal of micro-pollutants by activated carbon and ozone, the regeneration of spent carbon and the application of reverse osmosis as an alternative or complementary approach to physical-chemical treatment.

6. This paper gives a brief review of process selection and developments in reuse technology with special reference to current research activities and future research needs.

PROCESS DEVELOPMENT OF THE WINDHOEK PLANT

7. The demand for a safe water supply in the drought-stricken area of Windhoek prompted decision makers as early as 1960 to investigate the possibilities of effluent reuse. The sewage works at the time comprised conventional biofilters followed by maturation ponds. The quality of the HTE was rather poor particularly in terms of non-biodegradable detergents, ammoniacal nitrogen and dissolved organic substances. Retention of about 14 days in the ponds proved to be highly beneficial as regards improved microbiological quality and reduction by algal activity. The ponds also afforded some degree of quality equalisation and after diversion of abattoir and other industrial wastes a reasonably good quality pond effluent became available[5].

WRP I

8. The design of the original WRP I incorporated a dissolved air flotation process (DAF) for removal of flocculated algae (Fig. 1). Application of DAF proved to be technically feasible in WRP I and subsequently this technique was also success-

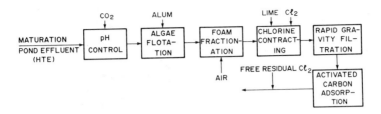

Fig. 1. Schematic flow diagram of WRP I

fully applied for the reclamation of other types of effluent[6]. The DAF process has gained popularity for various applications particularly to solve problems associated with the treatment of eutropied waters[7].

9. The DAF unit in WRP I was operated at a downward hydraulic velocity of 5.2 m/h at 200 m³/h and retention time of only 16 min which indicates the relatively small size of this particular unit[8].

10. Provision was also made for foam fractionation using a 5:1 volumetric ratio of air to water. This process achieved an equilibrium concentration of 1 mg/ℓ in terms of methylene blue active substances (MBAS) and prolonged the useful life of the activated carbon adsorption media. With the WRP I plant the spent carbon was discarded as no regeneration facility was provided for. Criteria for carbon replacement was based on a breakthrough of 0.5 mg/ℓ MBAS whereas COD values of up to 25 mg/ℓ were accepted.

11. During the first two years of operation the reclamation plant contributed an average of 13 % to the City's total water supply. Overloading of the sewage works however showed a deterioration of effluent quality and despite extensions to the biofilters, the chlorine demand for disinfection became economically unattractive. During 1972 it was decided to modify the Windhoek works to incorporate high lime treatment (HLT) and ammonia stripping (AS).

WRP II
12. In addition to HLT and AS, process modifications included additional activated carbon columns (Fig. 2). At a later stage on-site multi-hearth furnace regeneration of spent carbon was also introduced which favoured the overall treatment costs. These modifications gave much improved utilisation of the reclamation works but cost evaluations associated with maintenance of the stripping tower, fuel for CO_2 generation and power consumption proved to be high. It was evident that the HLT process was not an economically viable proposition for the Windhoek situation and also that more efficient biological pretreatment in the sewage works was called for, particularly to reduce ammoniacal nitrogen and dissolved organics.

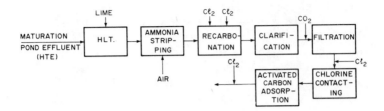

Fig. 2.   Schematic flow diagram of WRP II

WRP III

13.   Major extensions to the sewage treatment works in 1978 incorporated an activated sludge plant designed for biological nutrient removal.   Low ammonia concentrations could thereafter be achieved consistently and the reclamation plant was adapted for this improved quality by replacing HLT and AS with alum flocculation and interstage chlorination (Fig. 3).   However prechlorination had to be introduced to improve the settleability of flocculated algae using the clarifier which was specially designed for HLT.

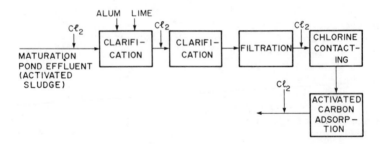

Fig. 3.   Schematic flow diagram of WRP III

14.   From the results in Fig. 4 it is evident that the activated sludge system ensured consistently low average ammonia values (1.9 mg/ℓ as N) and peaks were significantly reduced by the ponds.   This has greatly reduced the chlorine demand for complete disinfection.

15.   Nitrate removal in the activated sludge plant was, on average, most effective, with an average residual of 5.2 mg/ℓ (as N) over a 12-month period.   Concentrations exceeding 10 mg/ℓ occurred occasionally, but the maturation ponds effectively reduced nitrate to below 2 mg/ℓ .   The value of the ponds in terms of quality equalisation and improvement has been clearly demonstrated as being of great benefit to the overall process.

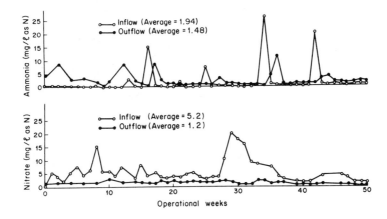

Fig. 4. Removal of nitrogen through the maturation ponds
(WRP III)

16. The average improvement in quality through the various
process stages is depicted in Fig. 5. Averages or orthophosphate (0.7 mg/ℓ), COD (7.3 mg/ℓ), nitrate (1.5 mg/ℓ, as N),
and ammonia (0.1 mg/ℓ, as N) in the final water confirmed the
effectiveness of the activated sludge cum pond system to produce an effluent suitable for reclamation by the WRP III
process configuration.

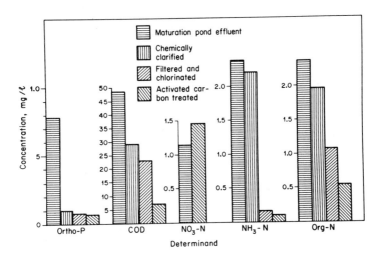

Fig. 5. Improvement of biochemical quality (WRP III).
(All values in mg/ℓ)

17.   The average increases in total dissolved solids
(75 mg/ℓ) was generally less favourable when compared with
the former HLT process but after admixture with the treated
dam water these values were well within specified limits for
potable water.

18.   During the WRP III runs the formation of trihalomethanes
(THM's) received special attention. Preliminary tests indicated that controlled prechlorination did not result in excessive formation of chlorinated hydrocarbons and facilitated
algae separation by settling. However backup research at the
NIWR demonstrated the need for algae removal prior to the
chlorination stage.

19.   Operation of the WRP III was greatly simplified in
comparison with the WRP I and WRP II process configurations.
Except for the activated carbon facilities WRP III is essentially similar to the conventional water treatment plant which
is situated adjacent to the reclamation works. The need for
activated carbon treatment in the conventional works has subsequently become necessary and additional activated carbon
columns were recently installed for this purpose.

20.   The WRP III mode of operation has been in use to date
and during the recent two-year drought period produced more
than 20 per cent of the City's water supply.

21.   Significant cost reduction could be achieved in the
WRP III mode of operation (Table 1). The main costs were
associated with the activated carbon treatment stage which
constitutes some 50 per cent of the running costs.

Table 1.   Comparison of treatment costs of the three modes
of reclamation (S A cents per kℓ, 1981)

	WRP I	WRP II	WRP III
Aluminium sulphate	3.33	–	3.96
Chlorine	1.29	8.25	3.21
Lime	0.24	3.34	0.47
Carbon	11.70*	11.70*	11.70
LP gas (CO_2)	3.51	3.24	–
Power	3.02	2.07	1.89
TOTAL	23.09	28.60	21.23

* Projected costs from WRP III

22.   A further process modification to reintroduce DAF as a
first-stage process is currently being considered. This has
become necessary because of sporadic carry-over of flocculated
algae despite prechlorination. This future modification is
expected to be a further safeguard against the formation of
THM's.

DEVELOPMENT OF THE STANDER WATER RECLAMATION PLANT

23. The SRP (capacity, 4500 m³/d) was originally designed to reclaim biofilter humus tank effluent (HTE) from the Daspoort sewage works in Pretoria. The process design necessitated HLT and AS to cope with the relatively poor and variable quality of the HTE. Several process modifications were made during the ten-year period from 1970 to 1980, and considerable effort was devoted to process optimization, operational control and minimization of production costs.

24. One of the major problems from the start as in the case of WRP I was the optimal control of breakpoint chlorination owing to diurnal variations in ammonia. A further aspect which demanded particular attention was the control of chemical stability associated with the high lime mode of treatment.

High lime treatment and ammonia stripping

25. Based on pilot-scale tests the use of HLT and AS was considered to be the best approach for reclaiming HTE and these processes were duly incorporated in the SRP (Fig. 6). During full-scale runs it was possible to reduce ammonia by 50 to 70 % for most of the time. Diurnal and seasonal fluctuations, however, still caused appreciable chlorination control problems which could be alleviated to some extent by the introduction of a quality equalisation basin[9]. The equalisation effect of the basin on pH and ammonia variability is demonstrated in Fig. 7.

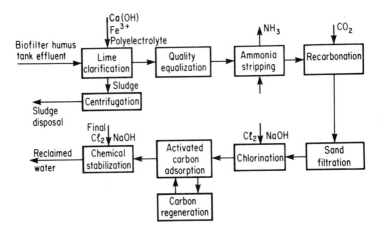

Fig. 6. Flow diagram of the 4500 m³/d Stander water reclamation plant (SRP)

26. The HLT process comprising two-stage clarification with interstage recarbonation had several advantages, such as conditioning of the water for ammonia stripping and disinfection properties at high pH. On the other hand, excessive scaling and maintenance of the tower proved to be disadvantageous. Despite these limitations the modified WRP II plant was designed

305

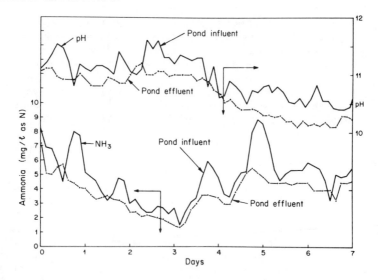

Fig. 7 Quality attenuation in the equalization basin (SRP)

to incorporate HLT and AS in order to achieve better overall utilization of the reclamation facilities for supplementing the urgent water demand at the time.

Activated carbon adsorption

27. At the SRP, provision was originally made for powdered carbon addition together with single-stage adsorption through granular beds but was discontinued when two-stage carbon adsorption with on-site regeneration were installed[10]. A detailed study of furnace operation in relation to physical properties and cost of regenerated carbon under South African conditions has been reported[11]. The multi-hearth regeneration furnace was eventually transferred to the Windhoek plant as part of the WRP II modification.

Ozonation

28. An ozone generator was introduced in 1977 to compare ozonation with chlorination for disinfection purposes. Results showed that chlorination was more economical, but ozonation had the advantage of extending activated carbon life. The phenomenon of biological activity on the adsorption media, has been investigated since the late seventies[12] and forms an important part of the current research programme.

Water quality

29. The intermittent operation of the SRP was routinely monitored for microbiological and chemical parameters as well as a wide spectrum of micropollutants. These results were reported[13] and, from a health point of view, proved to be satisfactory for all practical purposes.

30. Activated sludge effluent also became available on site during 1978 and test runs were carried out using ferric chloride as primary coagulant. This process modification was highly efficient in that an acceptable quality water could be produced without the operational problems associated with HLT.

31. Concurrently with process and operational investigations special attention was given to potable quality criteria for heavy metals, pathogenic organisms and the formation of volatile halogenated hydrocarbons[14] and a number of scientific dissertations on chlorination, adsorption, mole mass distribution of organics and kinetics of the calcium carbonate system and other in-depth studies were completed. A process design manual on water reclamation was also compiled[15] and this was complemented by a further design guide for planning of water reclamation schemes[16].

INTEGRATED PHYSICAL-CHEMICAL-BIOLOGICAL TREATMENT

32. A pilot plant in Pretoria for the treatment of up to 100 m³/d of screened raw domestic sewage was developed as an alternative approach to the reclamation of wastewaters. This facility was flexibly designed to enable various process configurations and operational modes to be compared. Extensive studies with HLT and DAF followed by biological treatment (LFB process) have been reported[17]. Later studies have included the use of ferric chloride as primary coagulant and partial denitrification was provided with a view to further quality[18] improvement and reduction of chemical treatment costs.

33. A larger scale pilot plant (300 m³/d) was also extensively tested at Athlone in the Cape area using the integrated physical-chemical-biological treatment approach.

LFB plant (Pretoria)

34. A flow diagram of the LFB plant is shown in Fig. 8 and the individual process units are briefly described below. The process stages may be classified into biological, chemical and polishing stages. The biological and chemical units are briefly described and typical performance data are given below.

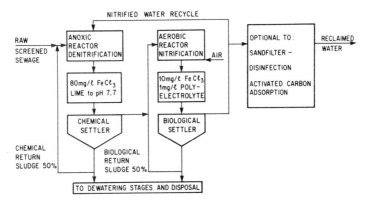

Fig. 8. Flow diagram : LFB plant system

35. Denitrification reactor. Raw screened sewage enters a
reactor vessel together with recycled streams from succeeding
stages. These include fully nitrified effluent together with
chemical, organic sludge. The recycled sludge serves as a
source of carbon for the purpose of biological denitrification.

36. Chemical clarifier. Ferric chloride is dosed to the
reactor outflow and lime added to control pH. Sludge is wasted
from this unit for separate processing.

37. Nitrification pond. An aeration pond and secondary
clarifier serve as an activated sludge process stage. Addition
of a further dosage of ferric chloride enhances the settleability of the activated sludge and secures a low effluent
turbidity for final polishing.

Operational results

38. The effluent from the activated sludge clarifier was
found to be consistently of a very high average quality in
terms of COD (<30 mg/ℓ), NH_3 (0.2 mg/ℓ as N) and soluble phosphate (<0.2 mg/ℓ as P). This quality proved to be excellent
for further polishing in that the chlorine demand for complete
disinfection was extremely low (<3 mg/ℓ). The low COD prior
to activated carbon polishing is also indicative of lower
activated carbon requirements as compared with effluents
derived from activated sludge or biofilter plants.

39. Spiking of the raw water with heavy metals also confirmed the safety barriers afforded by chemical pretreatment[19].
One disadvantage however is the approximately twofold larger
volume of sludge formed as a result of chemical dosage.

ATHLONE PILOT PLANT

40. Based on the abovementioned LFB process, a second pilot
plant (300 m³/d) was constructed and operated intermittently
over a period of some four years for the reclamation of secondary effluent derived from the Athlone sewage works near Cape
Town[18]. This particular sewage works (biofilter type) received
a stronger sewage which indicated wastewaters of industrial
origin. Its final effluent was of poor quality as reflected
by average residual COD values of 220 mg/ℓ, ammonia of 36 mg/ℓ
and nitrate below 1 mg/ℓ. The reclamation plant was designed
for selected industries such as textiles[20].

41. A flow diagram of the Athlone pilot plant is presented
in Fig. 9. HLT and AS were provided to remove the high ammonia
concentration partially. This was followed by a surface aerated activated sludge process with ferric chloride dosage to
improve secondary clarification. The polishing units comprised
prechlorination, gravity filtration, activated carbon adsorption and free residual chlorination.

Chemical and microbiological results

42. Representative data of chemical and microbiological
quality were consistently of a high standard. As in the case of
the Pretoria plant, extremely low ammonia levels could be maintained which enabled efficient disinfection control.

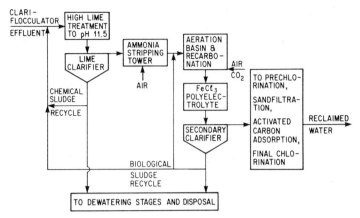

Fig. 9. Flow diagram : Athlone water reclamation plant

RUNNING COST

43. The cost of chemicals based on bulk supply prices and the electrical power costs (1979) for the Pretoria and Athlone plants were 6.7 and 13.8c/m³ respectively. The higher cost for the Athlone plant is essentially due to HLT and ammonia stripping power consumption[18].

CURRENT RESEARCH ACTIVITIES

44. During the extensive pilot and full-scale investigations many problems were encountered which received in-depth laboratory investigations.

45. A great deal of the current research is focussed on activated carbon because of its important role in water reclamation technology.

46. Because of the relatively high cost of activated carbon treatment, special attention is given to optimising regeneration conditions. An infrared conveyor-belt furnace is being evaluated[21] and utilized to study the effects of temperature and residence time on spent carbon intraparticle structure, and concomitant regeneration efficiencies. The infrared furnace has proved to be an effective regeneration system and an optimum operating region of 800°C 10/min to 850 °C/5 min has been identified. Temperature in excess of 850 °C must be avoided since excessive structural and pore volume distribution degradation is effected.

47. A test procedure for the degradation of carbon quality during regeneration has been developed[22]. Various carbon types, representing the base materials bituminous coal, anthracite, lignite, peat; and granular and extruded particle types are being evaluated.

48. Thermogravimetric analysis techniques are being developed[23] to serve as predictive tool concerning the regeneration of spent carbon, and for oxidation/adsorption process optimization. It has been possible to characterize the in-

fluence of oxidative pretreatment on the ultimate nature of the adsorbate associated with the spent carbon. A pre-ozonation dosage of about 3 mg/ℓ, for example, was found to afford a most suitable adsorbate thermal treatment response pattern.

49. A generalized technique for assessing abrasion resistance is being developed[24] which is applicable to all types of particulate activated carbons.

50. A comprehensive study is in progress, comparing the efficacy of oxidative pretreatment using ozone, chlorine, chlorine dioxide and oxygen, in promoting biological activity within the adsorbers and extending useful carbon life.

51. The occurrence, chemistry and control of compounds that act as precursors for the formation of organohalogen compounds upon chlorination, are being studied. Various physical-chemical processes (e.g. flocculation, oxidation) are being evaluated and optimized in respect to their efficiency in removing these precursor compounds.

52. Semi-pilot-scale studies are also conducted using reverse osmosis technology for water reclamation and special attention is given to pretreatment methods.

DISCUSSION AND CONCLUSIONS

53. The experience gained with reuse technology in South Africa has demonstrated the interdependence of process selection and design on the quality of the effluent which serves as a raw water intake to the reclamation facilities. This quality in turn is greatly dependant on the type of sewage treatment plant, and its operational performance in relation to climatic conditions and presence of industrial wastes, etc.

54. The need for maximum efficiency of the biological process stages cannot be over-emphasized particularly for the purpose of ammonia and disssolved organics removal. A great deal of time and effort was devoted in Windhoek and Pretoria to reclaiming a relatively poor quality HTE which necessitated provision for HLT and ammonia stripping. When these effluents were at a later stage upgraded by activated sludge treatment, the reclamation process at both Windhoek and Pretoria became technically and economically appreciably more attractive.

55. The DAF process is considered to be the best process selection for the separation of algae or flocculated organic substances as a first stage in a water reclamation plant.

56. The integration of physical-chemical and biological processes proved to have great potential for full-scale application and forms part of the current research programme. In addition semi pilot-scale investigations on reverse osmosis also show promise as an alternate approach which requires further research and development.

57. The importance of activated carbon treatment has initiated extensive ongoing research activities with great emphasis on the removal of micropollutants.

58. From a microbiological point of view reclaimed water produced during the various plant configurations at both

310

Windhoek and Pretoria consistently gave satisfactory results despite occasions where chlorination control was difficult and costly to achieve.

59. Water reclamation in South Africa is only considered for particular situations where large, expensive schemes can be delayed. It has the advantage of a standby facility rather than a continuous supplementary source.

PROGNOSIS

60. Water reclamation technology has been implemented, further developed and proven successful. Process refinements are continually being introduced, which favourably impact upon final water quality or process economics. As the quality of conventional surface and groundwater supplies tends to deteriorate in tandem with expanding industrialization, commercial and domestic activity, the distinction between water treatment and wastewater reclamation becomes superfluous. Advanced treatment technology and results from current research activities will therefore be increasingly applied in the wider context of all water supply systems.

ACKNOWLEDGEMENTS

61. Acknowledgement is due to all persons who participated in the past and those who are still actively involved in water reuse research. This paper is presented by approval of the Chief Director of the National Institute for Water Research.

REFERENCES

1. MORGAN W.S.G. Fish locomotor behaviour patterns as a monitoring tool. J. Wat. Pollut. Control Fed., Vol. 51, 1979, 580-589.
2. ISAÄCSON M. and SAYED A.R. Epidemiological studies on consumers of reclaimed water in Windhoek. Presented at Symposium on Health Aspects of Water Supplies, Pretoria, 1979.
3. NUPEN E.M. and HATTINGH W.H.J. Health aspects of reusing wastewater for potable purposes - South African experience. Presented at a Conference on Wastewater Reclamation Research Needs, University of Colorado, Boulder, 20-22 March 1975.
4. UNIVERSITY OF CAPE TOWN, CITY COUNCIL OF JOHANNESBURG and NATIONAL INSTITUTE FOR WATER RESEARCH. Theory, design and operation of nutrient removal activated sludge processes. Report prepared for Water Research Commission, Pretoria, South Africa, 1984, 1-1/12-6.
5. VAN VUUREN L.R.J., HENZEN M.R., STANDER G.J. and CLAYTON A.J. The full-scale reclamation of purified sewage effluent for the augmentation of the domestic supplies of the City of Windhoek. In Advances in Water Pollution Research, S.H. Jenkins (Ed.), Vol. 2, 1970, 1-32/1-39.
6. VAN VUUREN L.R.J., FUNKE J.W. and SMITH L.S. The full-scale refinement of purified sewage for unrestricted industrial use in the manufacture of fully-bleached Kraft-pulp and fine paper. In Advances in Water Pollution Research, S.H. Jenkins (Ed.), 1972, 627-636.

7.   VAN VUUREN L.R.J., PRINSLOO J. and DE WET F.J.   Pilot-
scale studies for the treatment of eutrophied water by dissolved air flotation filtration.   Presented at the 10th Federal
Convention of the Australian Water and Wastewater Association,
Sydney, 11-15 April 1983, 14-1/14-10.
8.   CLAYTON A.J., VAN VUUREN L.R.J. and ROUX B.   Development
of water reclamation technology in South Africa.   Wat. Sci.
Tech., Vol. 14, 1982, 339-353.
9.   VAN VLIET B.M., WIECHERS H.N.S. and HART O.O.   The efficacy of an equalization pond in a water reclamation system.
Prog. Water Technol., Vol. 9, 1977, 443-454.
10.   STANDER G.J., VAN VUUREN L.R.J. and DALTON G.L.   Current
status of research on wastewater reclamation in South Africa.
J. Inst. Wat. Pollut. Control, Vol. 70, No. 2, 1971, 213-222.
11.   VAN VLIET B.M.   Carbon regeneration in South Africa.
Prog. Water Technol., Vol. 10, Part 1/2, 1978, 555-563.
12.   VAN LEEUWEN J. and PRINSLOO J.   The effect of various
oxidants on the performance of activated carbon used in water
reclamation.   Presented at the IOA 5th World Congress, Berlin,
April 1981.
13.   PILKINGTON N. and VAN VUUREN L.R.J. Formation of trihalomethanes during chlorination of algae laden surface waters.
Presented at a Conference of the Australian Water and Wastewater Association, Perth, April 1981.
14.   VAN RENSBURG J.F.J.   Health aspects of organic substances
in South African waters - opinions and realities.   Presented
at a   NIWR-IWPC (southern African Branch) Symposium on Health
Aspects of Water Supplies, Pretoria, 15 Nov. 1979.
15.   NATIONAL INSTITUTE FOR WATER RESEARCH.   Manual for Water
Renovation and Reclamation, CSIR, Pretoria, 1978.
16.   MEIRING P.G.J. and PARTNERS.   A Guide for the planning,
design and implementation of a water reclamation scheme.
Report prepared for the Water Research Commission, Pretoria,
South Africa, 1-1/6-15.
17.   VAN VUUREN L.R.J., ROSS W.R. and PRINSLOO J.   The integration of wastewater treatment with water reclamation. Prog.
Water Technol., Vol. 9, no. 2, 1977, 455-466.
18.   VAN VUUREN L.R.J. and TALJARD MARK P.   The reclamation of
industrial/domestic wastewater.   Presented at 'Water Reuse -
from Research to Application', Washington DC, March 1979.
19.   SMITH R. and WIECHERS S.G.   Elimination of toxic metals
from wastewater by an integrated wastewater treatment/water
reclamation system.   Water S A, Vol. 7, no. 2, 1981, 65-70.
20.   TWORECK W.C., ROSS W.R. and VAN RENSBURG N.J.J.   The use
of reclaimed water in the textile industry.   Presented at the
Biennial Conference and Exhibition of the Institute of Water
Pollution Control (S A Branch), East London, 16-19 May 1983.
21.   VAN VLIET B.M. and VENTER L.   Infrared thermal regeneration of spent activated carbon from water reclamation.   To be
presented at the IAWPRC Biennial International Conference,
Amsterdam, 16-21 September 1984.

312

22. VAN VLIET B.M., VENTER L. and VENNEKENS M.J.A. Rapid regenerability assessment for particulate activated carbons. To be presented at the AWWA Water Reuse III Conference, San Diego, 26-31 August 1984.

23. VENNEKENS M.J.A. and VAN VLIET B.M. Adsorbate characterization by thermal analysis. Presented at the IOA/NIWR International Conference on Ozone and Activated Carbon, Pretoria, 26-29 March 1984.

24. VENNEKENS M.J.A. and VAN VLIET B.M. Aqueous suspension based abrasion resistance assessment for particulate activated carbons. To be presented as poster at the IAWPRC Biennial International Conference, Amsterdam, 16-21 September 1984.

Discussion on Papers 18 and 19

DR M. J. D. WHITE, Water Research Centre, Stevenage, UK
With regard to Paper 18 I would like more information on the
plant design described for 90% nitrogen removal, which was
presumably based on the South African work. The first
aeration stage with its high internal recycle could be termed
completely mixed, and in the Water Research Centre's work this
was associated with increased risk of sludge bulking. What is
the authors' experience? Could they expand on the working of
the second anoxic zone? What are the input levels of BOD and
nitrate? Is it supposed to be an endogenous respiration tank?
Do the authors have any experience of how well such a plant
works at low temperature, for example in an English winter,
and how difficult it is to operate?
   Taking the integrated approach what are the suggested uses,
or means of disposal, of the concentrate from reverse osmosis?
   Turning to Paper 19, many years ago the Water Research
Centre, and its predecessor bodies – the Water Research
Association and the Water Pollution Research Laboratory, did a
lot of work on advanced water and wastewater treatment
processes. This work stopped at the small pilot scale because
the needs of the UK did not justify the high costs of the
advanced processes. I am very impressed with the work carried
out in South Africa, presented in Paper 19, and I would like
to congratulate the authors and the South African authorities
for their achievement in demonstrating almost every advanced
treatment process on a realistic scale.
   As a process engineer I agree with one of the conclusions
that one must look at the overall system, as processes are
interdependent, and it makes sense to remove as much as
possible in the earlier, cheaper, biological processes to
reduce the load on later more expensive processes.
   Dr van Vuuren stated that the dissolved air flotation
process was best for the separation of algae and I wonder why
it was not used initially in Water Reclamation Plant
(WRP) III, and even now it is only being considered. Carry
over of algae from the settlement tanks at WRP III was
reported and I would like to ask whether the authors
considered, or tried, flocculation aids or weighting agents
such as polyelectrolytes or lime.

The authors highlighted the importance of biological activity in activated carbon columns, and enhancement using ozone. Have they considered biological activity on other carriers, such as sand, and does slow sand filtration (preceded by ozonation) offer a possible process within the advanced treatment train?

Dr van Vuuren mentioned the problem of diurnal flow. In terms of process selection have the authors considered splitting flow into a base load and peak flows, and having different processes in parallel, for example blanket clarifiers for the base load and flotation for peak flows?

The work on activated carbon is very important, particularly the criteria for exhaustion, the choice of optimum carbon type and conditions for regeneration. In view of the moving goalposts of the compounds to be removed by activated carbon, could Dr van Vuuren suggest a simple routine test for type of carbon and for carbon exhaustion?

MR H. G. GUNSON, Welsh Water Authority, Brecon, UK
In the course of the treatment of sewage effluent, especially with the addition of tertiary treatment, we produce more effluent, i.e. sludge. In view of the fully planned integrated approach to effluent reuse, would the authors of Paper 18 say what the way ahead is for the treatment and use of this sludge, so that maximum benefit can be obtained from it at the lowest cost?

DR J. D. SWANWICK, Sir M. MacDonald & Partners, Cambridge, UK
With regard to the subject of advanced treatment in the context of looking to the future I wish to say why Saudia Arabia wanted treatment to potable standards. I had been responsible for the treatment philosophy of Jeddah works to treat some 850 000 m^3/d, before the pilot scheme described by Dr Greenhalgh had been conceived. The requirement was for treated effluent to be piped back to households in a dual system to be used for toilets, floor washing and so on, and this requirement was confirmed against the advice of the consultants and WHO. It was therefore proposed, as a compromise, that treatment should be to international drinking water standards. This was accepted by the client and a scheme almost identical to that in the paper was developed. However, various installations failed because of inadequate operation and maintenance.

In looking to the future (and the Jeddah scheme was looking to year 2010), I am convinced that a client requesting advanced technology and having the financial resources to support it fully must expect to progress in that direction, provided the consultant fully explains the essential requirement for the appropriate level of engineering and the scientific experience and expertise to operate the system. Possibly the recent hiring of PhDs to run sewage works is an indication of moves in this direction, except that a high level of academic training is not a substitute for operational experience on large and sophisticated plants.

It might also be noted that the sequence of process stages involved in the Jeddah scheme was also likely to cost much less per unit volume of product water than reverse osmosis of seawater, which was being promoted at the time for the supply of drinking water and further illustrates the very high value of clean water in some regions. It also illustrates the enormous difference in the wealth of different developing countries and consultants need to respond across the whole spectrum with appropriate solutions.

MR J. M. PHILIPOT, OTV, Maisons Laffitte, France
In order to conserve the quality of our water resources more sophisticated treatment of urban wastewater has become a necessity. Purification processes using bacteria fixed on a fine granular particle bed are a means of obtaining increased purification efficiency for two reasons:
 - a biomass that is fixed to a supporting media reaches greater concentrations
 - the ability of a fixed biomass to settle is no longer an essential factor.
Fixed biomass systems therefore possess a purifying capacity superior to that of conventional systems.

Our overriding aims were to optimize the bacteria-fixing capacity (which depends on the type and granulometry of the media) and to optimize oxygen transfer (which depends on plant technology). The result of our research is the Biocarbone process (Fig. 1). This process consists of a downflow percolating filter and an immersed fixed bed into which air (or enriched oxygen) is blown at an intermediate level.

Technology of the Biocarbone aerobic filter
1. Filler or media
To be suitable for use in a Biocarbone unit the media must have a large specific area, a high degree of macroporosity and a low density and must be highly resistant to attrition. Among media offering interesting characteristics for the fixing of bacteria we find activated carbon, expansed shales or clay. Generally speaking, grain size must be 3-6 mm in order to reach a compromise between two conflicting requirements:
 - the size must be small enough to provide the maximum available surface on which the bacteria can attach, thus allowing maximum transfer of oxygen to the biological film
 - the size must be large enough to reduce the risk of clogging.

2. Oxygen transfer technology
Downflow percolation results in efficient oxygen transfer because of the counter-current flow of air and water. By injecting air at an intermediate level of the bed, the oxygen transfer in the aerobic bed is sufficient, at the same time allowing the suspended solids to penetrate deeply into the

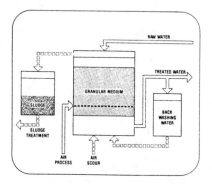

Fig. 1.   Biocarbone flow sheet

filtering media. The suspended solids are then entirely
screened by the non-aerated layer of particles which act as a
filter.

3. Removal of excess biomass
The filter clogs gradually owing to the growth of biomass and
the penetration of suspended solids. Regular washing runs are
required at intervals that will vary depending on the applied
load. The washwater is usually supplied from a treated-water
storage tank. The sludge resulting from the wash run is
recycled to the head of the plant.

Elimination of conventional pollutants found in domestic
effluent
The secondary treatment of a conventional effluent for applied
loads of 5-10 kg COD/m^3 per day gives a purified water with
COD = 45-55 mg/l, BOD = 10-15 mg/l and TSS = 4-10 mg/l. It is
also possible to treat industrial effluents such as brewery
and confectionery effluent.

Nitrification
The Biocarbone process can be used as a single-stage treatment
after primary clarification for the elimination of
carbonaceous pollutants and for nitrification. Nitrification
is obtained subject to two conditions:
- the carbonaceous load must be less than 4 kg/m^3 per day
- the nitrogen load applied must be less than 0.6 kg
  $N.NTK/m^3$ per day.
Under these conditions 80-90% of the influent nitrogen is
converted to reduced forms.
The Biocarbone process can be installed for tertiary
treatment downstream from a conventional treatment plant.

Treatment of drinking water
The Biocarbone process is suitable for treating surface water
(COD, $N-NH_4^+$) and is also an ideal choice for final treatment
after the denitrification process for drinking water.

Advantages of the Biocarbone process
The Biocarbone process, through the specific technique of
supplying air into an immersed biological bed at an
intermediate level, offers the following major advantages:
- very high effluent quality obtained in a single treatment
  stage, particularly as regards suspended solids
- the ability to operate at high loading rates
- compact installations.

MR C. BOWLER and DR S. H. GREENHALGH, Paper 18
Dr White will be aware of the work carried out in South Africa
by Professor G. V. R. Mavais at the University of Cape Town
and Dr J. L. Barnard in Pretoria which has been extensively
reported and which since 1981 has formed the basis of our
designs where nutrient removal is required.

Problems with sludges have been reported from several plants characterized by SVI's in the region of 150 ml/g and identified by filamentous organisms in the activated sludge. These problems have, generally, not been persistent and attributable directly to the flow conditions in the first aerobic zone. In fact, we would question whether this flow regime is really completely mixed and not approaching 'plug flow' - analogous to the conditions in a ditch system. It does appear most important to design the clarifiers as an integral part of the process.

The second anoxic zone is intended to enhance denitrification using the products of endogenous respiration as an energy source. We have found that a second anoxic zone is relatively slow acting compared with the first anoxic zone and is, therefore, incorporated in design only where the degree of nitrogen removal cannot be achieved. A problem in design is assessing the effect of varying load and flow conditions on a biological system and often a worst case steady state regime must be assumed.

We understand that plants of this type have been installed in Canada, Australia and South America and certainly the experience in Canada shows that satisfactory operation has been achieved at low temperatures.

We would like to thank Dr Swanwick for his interesting contribution giving the background to the Jeddah project. As he will appreciate, our Company as contractors became involved when the strategic thinking had been completed and the water quality objectives had been specified. We were faced with the problem of designing a plant to produce WHO potable standard water from sewage effluent and consider the scheme proposed to be the most suitable to achieve this end.

Whilst agreeing that the plant is a fairly complicated process and that prospective clients should be made aware of the operational requirements, our experience leads us to believe that the situation is not quite so daunting as might be imagined. Our company has constructed and supervised the operation of a similar plant in Saudi Arabia for five years. This includes similar process steps to Jeddah including a reverse osmosis stage. It has been our experience that the so-called 'high tech' reverse osmosis plant has much more reliability and is easier to operate than commonly imagined. In fact, it has been more reliable than the more conventional parts of the works. Blockages on lime lines and maintenance of chemical dosing and handling equipment absorbed more time than maintaining the reverse osmosis plant.

We would agree with Dr Swanwick's comments regarding water costs. As usual precise figures are not easy to obtain but many studies show the considerable difference in costs between treating brackish water and sea water by reverse osmosis. This is hardly surprising in view of the higher power requirements, lower recovery rates and more expensive capital plant associated with the sea water systems.

In reply to Dr Gunson, we would say that the consideration

which should be given to handling the sludge produced by reuse plants are no different from that to be given to any sludge. Both are part of integrated process systems in the widest sense. However, we have noticed an increasing tendency, particularly in the Middle East, to integrate the sludge handling into the whole treatment and reuse cycle. A good example of this can be found in Al Ain (in the UAE). At this plant the effluent receives tertiary filtration and is then used mainly for irrigation. The sludge is composted with municipal refuse and the compost spread on to the land which is under irrigation by the tertiary effluent. The technique has produced large areas of reclaimed land.

DR L. R. VAN VUUREN and DR B. M. VAN VLIET, Paper 19
Dissolved air flotation was not initially in (WRP) III because during the various plant modifications for high lime treatment in (WRP) II there was no need for DAF treatment. The sedimentation tank designed for (WRP) II was then used as part of (WRP) III. Currently this is being replaced by a DAF unit for more efficient algae removal. Flocculation aids such as polyelectrolytes have been tried as weighting agents but only with limited success.

Biological activity on sand with preozonation has not been studied and slow sand filter investigations have been limited to polluted surface waters. These processes may have potential within the advanced treatment train. Splitting flows into base load and peak flows has not been attempted. At Windhoek a constant flow is drawn to the reclamation plant with the excess discharged into a dry river bed.

It is considered important to subject a wide range of commercially available activated carbons to a screening evaluation in terms of abrasion resistance, apparent density and iodine number (e.g. AWWA Standard for Granular Activated Carbon B604-74) to narrow the field of potential candidates. Thereafter, isotherm and (at least) bench-scale adsorber breakthrough tests are indispensible in selecting a few suitable carbons. The criterion selected to identify exhaustion would depend on circumstances and local requirements. Local experience has shown that breakthrough of organohalogen precursors precedes exhaustion in terms of overall organics (DOC) removal, and precursor removal efficiency would be a conservative criterion.

General discussion

MR W. P. FIELD, Binnie & Partners, London, UK
The examples of the use of treated effluent for agriculture
given in Papers 6 and 7 are generally limited to relatively
small developments where either normal economic constraints do
not apply or the benefit is high for non-agricultural reasons,
for example city beautification. As far as larger-scale
commercial agriculture is concerned, only one example has been
shown, that of a wheat enterprise in Saudi Arabia, and here
again, as is well known, normal economic considerations are
inapplicable.

Paper 6 suggests that the large quantities of effluent from
major cities offer considerable potential for agricultural
reuse. In many countries improvements in efficiency of water
use for irrigation within the city supply catchment offer the
potential of comparable or greater savings in water without
the uncertainties associated with the health aspects.

In view of these factors, I would ask the panel's views on
the potential at this time for economic reuse of sewage
effluent in large-scale commercial agricultural development.

MR D. I. AIKMAN, Babtie Shaw & Morton, Preston, UK
The symposium has concentrated on direct reuse of sewage
effluent particularly in the field of agriculture. Irrigated
agriculture based on direct reuse is a minute part of the
world irrigation scene. Indirect reuse resulting from the
pollution of river supplies and of irrigation canal waters in
the developing countries of the world comprises much the
greatest area of reuse and I wonder if we are intentionally
applying double standards by ignoring this situation.

Much work has been done over the last 30 years in the USA,
in the reuse of effluent in agriculture using high technology
treatment systems. Can Professor Okun comment on the health
impact of such schemes?

DR A. ARAR, FAO, Rome, Italy
In arid and semi-arid areas like the Near East region
increased agricultural production, which is extremely
important, can mainly be achieved through irrigated
agriculture. However, the water resources of most of these

Reuse of sewage effluent. Thomas Telford Ltd, London, 1984

323

countries have been committed and there is a need to look for other, non-conventional, sources. Sewage water is one of the most important sources and serious considerations should be taken to bring it into use. Wastewater accounts for about 100% of water that could be used for irrigation in Kuwait, and the percentage is also very high for the Gulf states and many other countries in the area. For example, in Jordan about 30% of the total water resources available, and which will be totally committed by the year 2000, will be used for municipal and industrial purposes, and has to be utilized somehow or other to meet the demand of the increasing population.

Having also heard many of the aspects of treated sewage water, i.e. the chemical, health and technology of treatments, can the panel recommend a strategy for the treatment and use of sewage effluent with the main emphasis on agricultural usage? I was very much interested in the work in the UK on the recharge of underground water with almost untreated sewage effluent, without serious health and other pollution hazards. This process seems to be promising from the economic point of view and I would like to have guidelines for the suitable geological formation for such an exercise. Work in this field by Professor Brawer of the USDA Water Laboratory in Phoenix, Arizona, is also of interest.

With reference to the problem of night-soil, FAO has been researching in the field of organic recycling and the production of biogas and good manures from organic farm waste including night-soil. FAO has initiated a global programme in this field and many small demonstration projects on the production of biogas for heating, light and cooking have been set up in developing countries. A soil bulletin has been published by FAO on this subject entitled Organic Recycling.

MR H. M. J. SCHELTINGA, Ministry of Housing, Physical Planning and Environment, Arnhem, The Netherlands
In the framework of the concerted action Cost 168, from the Commission of the European Communities, much attention is given to the hygienic aspects of the use of sewage sludge in agriculture. There will be a similarity between these aspects of sewage sludge and those of sewage effluents discussed at this symposium.

PROFESSOR D. A. OKUN, Panel speaker
In response to Mr Field, the greatest opportunities for realizing significant economic and resource benefits from reuse of sewage effluents will be for non-potable use in urban settings, where the reclaimed water can replace limited supplies of high quality drinking water, so that these minor contribution to large-scale commercial agriculture development.

With regard to Mr Aikman's query about health risks, the water quality requirements for reuse in agriculture depend upon the nature of the agricultural enterprise. Where exposure to the irrigation and the crop is small, treatment requirements are minimal. Where spray irrigation is used and the crops are to be eaten raw, the quality requirements will be rigorous. But even for the latter instance, high-technology systems are not required. The filtration and disinfection of a secondary effluent should be adequate.

MR J. P COWAN, Panel speaker

The observations made are generally valid but the developments in agriculture in Kuwait whereby fodder is grown to support a thriving dairy industry and the fattening of sheep for local consumption are not insignificant when compared to the amount of agriculture in existence in Kuwait a decade ago. In addition it should be recognized that the water resources of Kuwait are scarce and the use of effluent is economically attractive when compared to the cost of desalinated water. Before any scheme is implemented it should be subject to an overall economic assessment. But until integrated policies are formed and respective government bodies are co-ordinated, it is likely that conflict between water, sanitation and agriculture priorities will continue.

It is true that large quantities of potable water are lost in water distribution systems and governments are increasingly becoming aware of the benfits of spending money on leak detection surveys and upgrading their water distribution systems, thereby conserving water or making more available for distribution. In view of the high cost of the direct use of treated sewage effluent for agriculture the most likely extensive reuse of sewage effluent in large commercial agricultural projects is where the possibility of dilution by adjacent rivers exists.

Closing address

Professor P. C. G. ISAAC, Consultant, Watson Hawksley, UK

It is clear from the papers and discussion that the reuse of
sewage effluent, in one form or another, is common in all
parts of the world - even where, in fact, there is no
waterborne sewerage. Nevertheless, this symposium has not
been teaching grandmothers to suck eggs, but has been rather a
sharing of experience. I must try to pick out the highlights.

I have long been conscious that we public health engineers
suffer from the mid-19th century myth that money can be made
from waste; it cannot! With this in mind it is important to
remember that the principal object of sewage treatment is to
allow human and industrial effluents to be disposed of without
danger to human health or unacceptable damage to the natural
environment. The additional cost that can properly be
incurred in reusing sewage effluent must be governed by the
value of the effluent in very local and specific
circumstances. The importance of multidisciplinary
collaboration cannot be overemphasized.

A further thought that a reading of the papers gives me is
that we do not have adequate methods for measuring possible
deleterious health impacts against more easily demonstrable
advantages. Too often engineers are ignorant of possible
health damage or too ready to give way to the pressure of
those who suggest that, as it becomes possible to measure
nanogram quantities of substances known to be dangerous at
much higher concentrations, so we should remove such material
as a matter of course. We must be prepared to put up
consciously the same kind of argument for factors of safety as
we accept in, for example, structural design, a point well
made by Mr Perret in the discussion on Paper 17. For example,
I wondered, when reading the splendidly detailed Paper 12 by
Dr Avendt, whether the use of very highly treated water could
be justified, in other than very special circumstances, for
the recharge of groundwater. On the other hand it is
essential, in the indirect reuse of effluent in lowland
waters, to ensure that accidental, or indeed deliberate,
pollution does not result in disaster. It sometimes seems
that the factor of safety is small here; direct pumping via
treatment into potable supply is hardly satisfactory.

Professor Diamant, in Paper 2, makes the point that good water is a rare thing in many developing countries, but he is less than fair to consultants in his comments on their design of systems. Those of us who have worked in many parts of the world know the extreme pressures, resulting from demands of prestige and politics, that can dislocate the most intelligent planning by the overseas expert; the problem is often much nearer home than he recognizes. It is valuable that more and more engineers have not only overseas experience, but also a feeling for the medical aspects of the work like, for example, Dr Feachem and Professor Pescod.

The papers from Australia, South Africa (where effluent has been reused for a very demanding purpose at Windhoek), the US, the Middle East and the Netherlands have been most helpful in demonstrating both the common problems and the large measure of common approach.

In much of the Middle East water is very short, but relative prosperity calls not only for waterborne sanitation, with its high load on the water resources, but also for an increasingly beautiful environment with trees, parks and flowers. Amenity reuse of water is, therefore, likely to have a high priority in the area. Fortunately the health requirements are perhaps lower than in other applications.

As Mr Squires shows in Paper 4, effluent is not strictly reusable until it has been discharged; recycling in a single industrial complex is, however, probably the most effective method of minimizing impact on water resources, and ought to be considered even before reuse. The detailed information given in his paper will be widely useful, as is the observation of Dr Taylor in Paper 5 that, in the schemes of industrial reuse with which he has been concerned, the cost of reclaimed water was of the order of 10-25% of that of the fresh mains water it replaced.

The principal worldwide reuse of sewage effluent, however, will be for agriculture, and it is necessary to reconcile the topically urgent need for more food with the possibility of spreading disease by this means. Papers 6, 7, 15 and 16 provide valuable criteria by which we can judge the value of agricultural reuse in any particular case. The balance of advantage must always rest with the intelligent reuse of effluent for raising food. As Dr Huggins (Paper 8) and Dr Payne (Paper 9) demonstrate, fish farming and aquaculture are making their contribution - in some parts of the world, indeed, with the minimum of preliminary treatment for the sewage. While I was vice-chairman of the board of trustees of the Asian Institute of Technology I was privileged to see the excellent research into aquaculture being carried out at the Institute, with considerable support from the Overseas Development Administration. It offered great hope in a country where fish is a main source of protein.

It is perhaps no more than implicit in what has been said at this symposium that the spread of microbiological disease can be prevented, but that the significance and prevention of heavy metals and other micropollutants are problems requiring urgent attention.

Mr Scheltinga emphasized in Paper 10 that, for certain amenity reuses, the quality standards are very demanding. In a sense we are here reverting to the kind of standards for disposal examined so exhaustively by the Royal Commission between 1898 and 1915.

In the UK, at present, little effluent is used for groundwater recharge, and Paper 12 by Dr Avendt can usefully be compared with that of Dr Montgomery and his co-authors (Paper 14). Mr Fougeirol's contribution from France was most interesting as it showed the direction being taken in that country. In this field the considerable research being pursued into the attenuation of liquor leached from landfills has much to tell us of the treatment capacity of geological strata.

Is it possible to envisage the increasing use of disinfected effluent to replace good water in low level uses such as lavatory flushing? As an example it has been proposed by Watson Hawksley to use disinfected and dyed secondary effluent from the Sha Tin works for this purpose, instead of the seawater which is used in parts of Kowloon, Hong Kong.

Mr Porter and Mrs Fisher (Paper 17) have taken a brief look at the economic assessment of schemes of reuse, emphasizing that economic and financial appraisals may give different results. Papers 18 and 19 have taken us to the threshold of today's available techniques in treating sewage effluents for the full range of reuse possibilities, as did that of Dr Avendt on Cedar Creek (Paper 12).

If I am required to finish by crystal gazing, I must say that I see the most immediate demands as philosophical rather than technical. We badly need the means of comparing, with the same weight, the quantifiable advantages of a particular scheme of reuse and the possible deleterious impacts of that scheme. By the same token we need to place the possible health dangers of heavy metals and other trace constituents in effluents in context with the other incidents of life - and here one remembers that Mr Cowan (Paper 7) suggested that it was at least as dangerous crossing the road to the central reservation as it was to sit on the possibly polluted grass! Where underground aquifers are to be recharged we need to know more of the purifying capacity of geological strata. And at all times we must give at least as much attention to preventing the misuse of water as to its recovery when it has been used and discharged.

ADDENDUM

MR J. P. COWAN and MR P. R. JOHNSON, Paper 7
In reply to Dr Butler, the merits of placing greater emphasis
on the selection of crops that will undergo a manufacturing
process and so remove the need for tertiary treatment poses
many questions without clear answers.

One of the greatest problems is the general lack of a
comprehensive policy for water, sewage and effluent
utilization in particular. This is not surprising since in
many countries effluent utilization is a new concept and many
of the bodies responsible for the management of water and
sewage are still in their infancy. Where departments are well
established, co-ordination between them often poses
difficulties and the prospect of a single responsible body to
establish policy sometimes appears remote. Simple, well tried
technology applied by new departments to grow what they desire
and understand is more likely to bring the most success.
Satisfaction with this success will encourage expansion and
development but in the Third World it is unlikely that the
production of those crops from which pharmaceutical products
can be derived will precede the production of more traditional
crops.

The research currently being undertaken at Portsmouth
Polytechnic is one of the new approaches referred to by
Professor Pescod and is welcomed. The developments in
association with the University of Florida will be watched
with interest.

Regarding Professor Diamant's point, we endorse the view
that all crop production involving sewage effluent should not
be undertaken in an informal manner and that careful
regulation and supervision is essential to safeguard the
health of both operatives and consumers.

The hazard of heavy metals in sewage effluents, mentioned by
Dr Arar, must not be overlooked. However, due to the absence
of established industries in the Middle East, sewage has been
predominantly of a domestic nature and no significant problem
has arisen. As industry grows, bodies responsible for
effluent projects must be vigilant to ensure that such hazards
do not arise. If the need does arise, trade effluent
discharge policies will have to be established and
implemented. These potential problems highlight the need for
close co-operation or integration of water, sewage, municipal
and agricultural departments.

On the absence of significant groundwater recharge in the
Middle East, this matter was not referred to in Paper 7.
However, in Qatar the recharge of the northern aquifer has
been considered but to date effluent availability and economic
grounds have precluded its development. In Taif, Saudi
Arabia, a large-scale experiment in which treated effluent is
mixed with groundwater via infiltration basins is in hand.
The cost of such schemes is not insignificant and the
consequences are not fully predictable. Where there is a risk

of the groundwater being extracted for potable use, we consider that further advanced treatment of the sewage effluent prior to recharge is essential.

The salinity of soil has not posed a significant problem on schemes with which we have been associated, but due consideration must be given to type of crop, salt tolerance and other factors mentioned by Dr Arar.

The major problem encountered is the contamination of raw sewage by the infiltration of saline groundwater, notably in the Gulf States where sabkah is extensive. On some occasions this has led to the salinity of the treated effluent being too high for agricultural applications. Where this has occurred, effluent has been discarded until the source of infiltration has been traced and measures taken to prevent contamination of the untreated sewage.

With regard to Mr Angier's question, we are unaware of any major problems with clogging but it should be noted that much of the wastewater available to date has been utilized by the flood and furrow method of irrigation. The probability of long-term clogging has been minimized on our project by the treatment of the wastewater to tertiary standards, and no specific additional filtration is envisaged.

In the early days of effluent use in Abu Dhabi there were difficulties in enticing people away from green areas since they were 'oases' in the middle of a desert. More recently the provision of parks with walkways, seats and recreation areas together with strategic fencing has led to a noticeable decline in the number of people sitting on grassed areas. This has also been assisted by an education programme by the municipality who, in addition to restricting access to the parks, have erected signs warning that the water used for irrigation is not pure. Health risks arising from public contact with effluent irrigation are clearly associated with the water-borne diseases endemic in the population served by the sewerage network. However, there is little evidence of significant transmission of disease through well-operated treatment facilities.

In answer to Mr Ellis, there is little evidence that sunlight is a major factor in disinfection and any effect probably only penetrates a thin layer. Time can be a major factor.

The evidence for effective removal of protozan cysts and worm eggs by tertiary treatment is provided in technical report No. TR128 'The fate of pathogenic micro-organisms during wastewater treatment and disposal' which is published by the Water Research Centre. This refers to the effective removal of helminths by both sand filters and microstrainers.

The source of Table 3 in Paper 7 was 'Appropriate technology for water supply and sanitation' by Feachem, Bradley, Garelich and Marci (1980), World Bank, Washington, chapter 1, pp. 138 and 139; see also 'The public health engineer, January 1983, p. 37.

10/10 is a typical performance standard adopted for tertiary

treated effluent and is accepted as such by most national and
local authorities. Despite various operational difficulties,
chlorination remains the most cost effective method of
disinfection and can be applied with confidence to a tertiary
treated effluent. It is accepted that chlorination will not
prevent the regrowth of bacteria with the passage of time, but
without incurring a very considerable cost penalty we are
unaware of a more suitable process.

In answer to Mr Farrer, the schemes described in our paper
are all, to a greater or lesser degree, in their infancy since
they are the tail end of sanitation projects conceived within
the last decade or so. Some schemes with which John Taylor
and Sons has been associated have included provisions to
permit the augmentation of supplies by potable water, but this
has rarely been implemented, not least because the periods of
maximum demand for effluent and potable water coincide. The
feasibility of long-term effluent storage has been considered
on several projects but has been found uneconomic even when
examined in the context of substitution for high cost water
produced by MSF distillation or reverse osmosis plants. Long-
term effluent storage such as aquifer recharge has been
considered in some locations but has not been implemented yet.
Surplus effluent in Doha has been discharged to a remote wadi
where a large lagoon with associated wild life has formed. In
due course it is anticipated that much of this water will be
used as the infrastructure develops and the financial
resources permit.

With reference to Mr Hennessy's comments, while wastewater
has been utilized in one way or another for centuries, it is
only relatively recently that treated wastewater has been
available for direct use. Wastewater has not been used
extensively in the Third World simply because capital costs
have been prohibitive and also because a trained local
workforce to operate and manage such schemes is not readily
available. Furthermore, treated wastewater is a resource only
recently available to the more wealthy states of the Middle
East. The first priority for Third World cities is a good
water supply followed closely by an adequate sanitation
scheme.

The suggestion of unified management to cover sewage
treatment and irrigated areas is attractive but the difference
in disciplines involved is likely to hinder this. It is of
note that even in some developed countries water and sewage do
not come under the same umbrella. As the use of treated
wastewater becomes more widespread, the need for government
bodies empowered with appropriate legislation to co-ordinate
and even enforce policies becomes greater. However, progress
towards the establishment of joint bodies responsible for
water, sewage and wastewater application (including municipal
applications and farming) has been slow and we do not envisage
that this will change in the immediate future.